'5⁰⁰

D1442579

Mapmakers of America

ALSO BY S. CARL HIRSCH

Cities Are People

On Course! Navigating in Sea, Air, and Space

Printing from a Stone: The Story of Lithography

The Living Community: A Venture into Ecology

Fourscore . . . and More: The Life Span of Man

This Is Automation

The Globe for the Space Age

S. Carl Hirsch

Mapmakers of America

*From the Age of Discovery
to the Space Era*

Illustrated by William Steinel

THE VIKING PRESS   NEW YORK

For the Argentos
—Joseph, Inga, Vance, and David—
who cherish our land
and its westering pathfinders.

First Edition

Copyright © 1970 by S. Carl Hirsch

First published in 1970 by The Viking Press, Inc.,
625 Madison Avenue, New York, N.Y. 10022

Published simultaneously in Canada by
The Macmillan Company of Canada Limited

Library of Congress catalog card number: 70–102922

526.8   Cartography
910.9   History of Geography—Discovery and Exploration

Printed in U.S.A.

Trade 670–45439–7   VLB 670–45440–0
1   2   3   4   5   74   73   72   71   70

Acknowledgments

My special thanks for a careful rechecking of the facts in this book go to J. O. Kilmartin, Chief, Map Information Office, Geological Survey, United States Department of the Interior.

Others who helped greatly in reading and correcting portions of the text are: Wellman Chamberlin, Chief Cartographer, National Geographic Society; Herman R. Friis, Director, Center for Polar Archives, National Archives and Records Service, General Services Administration; Bruce C. Ogilvie, Geographer, Rand McNally & Co.; Patrick D. McLaughlin, Archivist, National Archives and Records Service, General Services Administration.

Still others who assisted in many ways were: Donald L. Zylstra, Public Affairs Officer for Space Science and Applications, NASA; Nordis Felland, Librarian, American Geographical Society; Morris M. Thompson, Assistant Chief Topographic Engineer for Research and Technical Standards, Geological Survey, United States Department of the Interior; Frederick J. Doyle, Chief Scientist, Raytheon Company; Walter W. Ristow, Chief, Geography and Map Division, The Library of Congress; Athos D. Grazzini, Associate Chief Cartographer, National Geographic Society; A. P. Muntz, Chief, Cartographic Branch, National Archives and Records Service, General Services Administration.

My deepest gratitude goes to my wife, Stina, who helped rummage the archives for map lore and who traveled with me on the mapmakers' trails.

Contents

Mapmakers of America

1.
A Restless Quest

In the spring of 1968 a camera in a Gemini space capsule took a series of photographs of the entire southwest region of the United States. Startling in its detail and broad in scope, a photomap made from these pictures revealed how far we had come in the mapping of this country.

In the very center of the Gemini photomap was the ancient land of *pueblos*, where countless generations of Indians have built their many-leveled shelters against the region's fiery sun and slashing windstorms.

White men, Spaniards, first arrived here in the year 1540. At this place the explorer Coronado completed what was probably the first important map of inland America. It was a map that marked out his six-month journey deep into what is now the Southwest of the United States.

But this was more than a record of land crossed and observed. The map was a story of a restless quest. In its lines and symbols Coronado revealed how he and his followers, half crazed with hunger and thirst, stalked a golden dream. It was here they came to seek the fabled Seven Cities of Cíbola.

In the chill February dawn, a strange caravan moved out into the wilderness. Its colors blazed against the drab Mexican landscape.

Fluttering banners led the column. Two hundred young noblemen pranced their fine horses. Their plumed helmets and glinting armor caught the morning sun.

Behind came the Indians in brightly marked blankets, painted headbands, and parrot feathers. In the rear marched the slaves in bold-colored African dress, herding the pack animals.

At the head rode Francisco Vasquez de Coronado, a proud young man in a golden helmet, mounted on an armored horse. As he chatted amiably with his companions one might have heard in his voice the ring of adventure, the promise of glory.

Like the *caballeros* with him, Coronado was a young Spanish nobleman bent on overcoming the misfortune under which he was born. By Old World custom and law the large fortunes and estates were handed down from father to eldest son. But to every younger son of each noble family went only a sword and a blessing, and the wish that he might someday find wealth and fame on his own.

The choices were few. The young man might join those thronging about the royal court at Madrid with the hope of being favored with some rich post. Or he could enter the mighty

Spanish army or navy with the dream of becoming a hero. Or else he might venture as a *conquistador* to the farthest frontier of Spain's lands abroad, with the hope of finding—who knows? A small empire of his own where he could live in ease and luxury? A rich new trade route to the Orient? A treasure-filled paradise in the wilderness?

In the early sixteenth century hundreds of such young men had followed the Spanish conquerors into the Indian towns of Mexico. Coronado, himself a second son of a noble Spanish family, had gathered around him a band of eager *caballeros*. To them he whispered a story which he had been told.

Far to the north, Coronado said, there was a group of Indian cities—seven, to be exact. These cities were rumored to contain riches beyond the wildest dream of kings. The towering temples were filled with priceless jewels, their domes inlaid with matchless turquoise and plated with gold.

The journey, Coronado warned, was long and perilous. But to young, hardy men, filled with ambition, this seemed the way to immeasurable wealth. "Will you join me?" asked Coronado. And two hundred willing voices answered, "*Sí!*"

At that moment there were already dozens of Spanish expeditions pursuing visions of glory in the New World. Such a quest might have begun with a story wrung from a wandering Indian youth, captured and brought before a Spanish overlord.

"That gold charm hanging about your neck—where did you get it?" he asked the terrified native boy. "Speak now or you will never speak again. Tell us where the gold is hidden or your eyes will never again see the glitter of gold!"

With a dagger pressed against his throat, the frightened captive knew that the only way to save his life was to tell as big a lie as he could invent.

Gold, yes, plentiful nuggets overflowing the vaults and chests.

Jewels—as big as plums! Where? Twenty days' journey—or perhaps twenty-five. Northward—and a little east. Or was it west?

Soon such a rumor, circling about, returned like an echo from the distant mountains. It came back louder each time until it seemed to carry the trumpeting sound of the truth. The Indians had at last learned how to get rid of these greedy newcomers whom they could not defeat in battle. Were the Spaniards plundering their village? Were they eating the last stocks of maize and dried meat? Were they abusing the women and making slaves of the Indian children? Then tell them a story of gold beyond the hills—and soon they would be gone!

To Coronado, it mattered not that the Indians he saw coming down from the north country lived even more poorly and humbly than those of Mexico. The belief in the Seven Cities of Cíbola was so strong that these cities had already appeared on Spanish maps. And Coronado was determined to prepare a map of his own which would show every inch of the route to the treasure trove.

The harsh winter turned into a rainy spring, with Coronado's corps of younger sons making its way northward roughly parallel to the Mexican west coast. One evening, as the sun hung like a gold piece in the western sky, Coronado raised his hand and called the halt for the day.

The little army spread itself out in a pine grove. They had long since become weary of the journey. Food supplies were now gone, and the barren land offered little game or edible plants.

At nightfall the commander set up his drawing board under a flaring pine-pitch torch. Carefully he sketched the landmarks along the route of the day's travel: the long range of reddish mountains to the west and its central peak, the sandy treeless

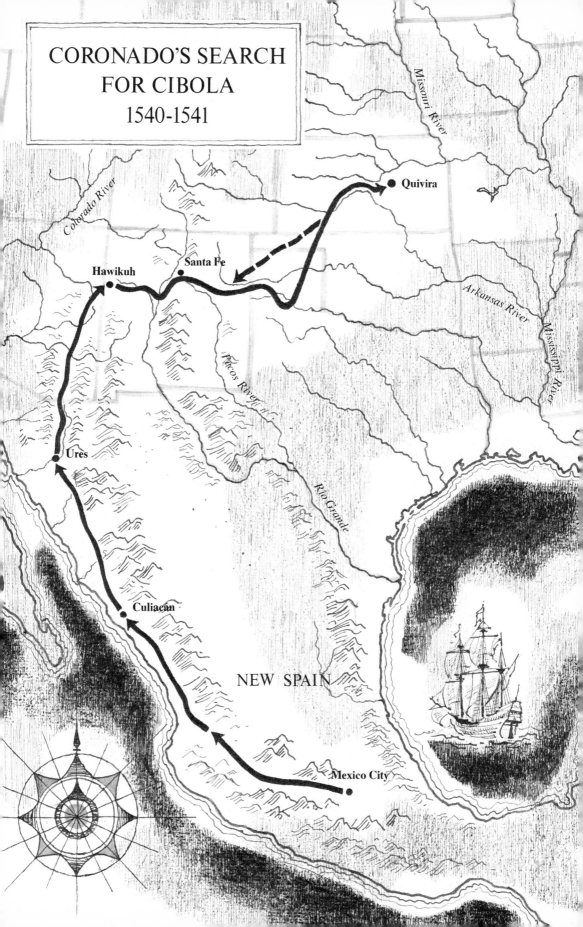

CORONADO'S SEARCH FOR CIBOLA
1540-1541

Missouri River

Quivira

Colorado River

Santa Fe

Hawikuh

Arkansas River

Mississippi River

Pecos River

Ures

Rio Grande

Culiacán

NEW SPAIN

Mexico City

flats just ahead, the place where they had crossed a dry river bed and found the whitened bones of long-dead animals. All these were set down on Coronado's map.

That night, while most of the fatigued band went to bed hungry, a few of the men went foraging for food. They found a patch of plants with pulpy roots, somewhat like turnips. The famished men downed this food raw and fed some to the horses. By morning, horses and men were dead.

The procession moved out of that place quickly, leaving their poisoned companions in trailside graves. Coronado had hoped to find here the strength and courage to cross the wasteland ahead. But now drought and hunger and disease stalked the expedition. The worst days were yet to come.

Two thousand miles and two hundred days before, this had started out as a high-spirited treasure hunt. Now it was only the zeal of the leader that pressed them on. Coronado had already sketched in on his map the place where he expected to find the Seven Cities of Cíbola, and he kept his eyes fixed on that glimmering goal.

The landscape had turned into a furnace. Baked by the midsummer sun and parched by the hot wind, the exhausted party dragged themselves through the craggy canyons and up the sandstone hills.

Now it was a trail of death, haunted by buzzards and coyotes. Coronado pushed onward, his face blackened by the sun, his lips blistered, his eyes strained toward the distant horizon. His army now followed like men pursuing a mirage across the desert.

It was late in a July day when a forward scout came running back with a cry that he could scarcely utter out of his parched throat.

"Cíbola," he cried, "just beyond the bluff!"

Now the ragged little army, shaking with thirst, hunger, and weariness, struggled up the last hill. They gasped at what they saw below in the sunset. Gold! Gold everywhere! Here at last were the shimmering cities of their dreams, tall structures of blinding radiance. Here was glittering glory!

But the hot sun had begun to drop below the rim of the desert horizon. And now suddenly the landscape was being transformed. The dazzling cities were changing before their very eyes!

What had appeared pure gold in the sun's brilliant rays turned into red clay. What Coronado saw below him now was a modest Indian village, its high-rising pueblos built of mud and stone, its people living their simple lives.

This was Cíbola.

Coronado's story is typical of the great adventure that unfolded in the mapping of the wilderness which was to become the United States. Men began with a dream of what they might find in this unexplored and mysterious country. Curious and courageous, they wandered on until they either proved or disproved their own beliefs.

Coronado went on into the center of Kansas without finding the treasure he sought. His expedition was a deep thrust by white men into what is now the United States. But he considered it a failure. He could not have known that the country he explored would produce more wealth than the greatest of the Old World empires. Nor could he have guessed that here democracy would someday challenge the age-old rule of kings.

Least of all did he understand that on his map he had traced one long line which would be part of a growing portrait of America. Mapmakers to come would add lines of their own until the likeness of this land would take shape.

But a truthful picture would not come easily. Men would add to the map things which they only imagined and never saw. On this map would appear swirls and squiggles, pothooks and chicken tracks representing not facts but fancy. False lines would be scribbled and garbled, some to be corrected only in the course of centuries. There would appear on this map rivers that never ran and seas that were dry. Mountains would be traced across places where there was really nothing but plain. The outline of the land would appear like something seen through a twisted glass, distorted and faulty.

To the European newcomers the New World was a blank, with a huge question mark scrawled over it. But men are uneasy with their own ignorance. Some stirred themselves to truth-seeking and exploration; others were content to fill the void with something out of their own imagination or rumors they had heard.

In a million square miles of land he had never seen, the imaginative man had room enough to build his own fantasies. He reported on Indians with long beards and Indians who spoke Welsh. The flatlands, he claimed, abounded with Peruvian llamas and giant creatures of the elephant family. In the woods he supposedly battled with leopards and tigers. He told of rich diamond mines, and he repeated the tales of cities bursting with treasure long after Coronado had learned sadly that they were not true.

The most stubborn falsehoods were those that told about the shape of the land masses. Maps were drawn which flatly contradicted each other. Was North America an island, a string of ocean isles, or a barrier reef? Was it part of an Oriental continent—or an island off either the Asian or the South American coast? A map could be found that supported any one of these ideas.

From the very beginning the map of the New World was a kind of arena where truth grappled with falsehood. One skirmish took place on the ship in which Columbus returned home from his voyage of discovery. The admiral exploded in anger when he looked at the first sketches made by his mapmaker, Juan de la Cosa. The drawings showed the West Indies islands visited by the expedition. No, roared Columbus, Cuba was not an island! Cuba was part of the Chinese mainland. And Columbus forced Juan de la Cosa to sign a statement that this was the truth.

Early in the year 1524 a lone ship emerged from the wintry fog off the coast of North Carolina. From its mainmast fluttered the blue flag of France, and on its stern was the name *Dauphine*.

The captain was Giovanni da Verrazano, a Florentine who sailed for the French because his own country was unable to send expeditions to the New World. This seafarer and his brother Gerolamo, a distinguished mapmaker, were to influence the map and the image of America for generations to come.

Like others who explored the new continents, Verrazano fixed his eyes on the riches of the Orient. He saw America as a barrier, thinking only of how he could go beyond it or around it or through it. Verrazano probed the long eastern coastline, looking for an opening.

Great welcoming bonfires appeared along the sandy beaches, and the explorer went ashore. He studied closely the features of the American Indians and thought he saw in them the face of the Orient. "They are like to the people of the east parts of the world," he reported to the French king, "especially to them of the uttermost parts of China."

To Verrazano, everything he saw seemed further proof of what became the most enduring myth of America—that somewhere was a waterway through the wilderness, leading to the Far East. He sailed his graceful caravel into every promising inlet from Cape Fear to New York Bay.

Not far west of the seacoast Verrazano seemed to see a broad expanse of water which he believed to be "the Oriental sea."

For the next one hundred years many maps of America pictured the "Sea of Verrazano," indicating that anyone who found it could sail on to Asia. But no one did.

In that century after Columbus many voyagers came to North America. Most of them approached it gingerly as though it were some strange beast cast up by the sea. They skirted the Atlantic and Pacific shores, ventured into the river mouths, sketched the shoreline, and guessed at what lay beyond.

Early mapmakers showed a long, massive mountain range stretching its summits east and west across the middle of the continent. We now know that part of the United States as the broad valley of the Ohio River.

In 1651 a map showed the Hudson River as running to "the Sea of China and the Indies." Shortly afterward a Maryland map with an explanatory text pointed out that the Allegheny Mountains, actually located in the eastern United States, were "the middle ridge" of North America.

Far into the nineteenth century, maps of America's West showed a "Buenaventura River" cutting a broad valley across the central region on its way to the sea. The river was pure make-believe.

There were, of course, bluffers and yarn spinners who invented some of these fables, as well as outright liars. But others were well-meaning explorers who didn't understand what they were looking at. Among these was a nineteenth-century Army officer named Zebulon Pike, whose name is attached to a famous Colorado peak which he never climbed. It was said of Pike that he discovered more by accident than he did on purpose. Pike seemed to be forever traveling in the wrong direction, exploring the right fork of the wrong river, blundering into forbidden Spanish territory. He mistook the Missouri River for the Platte and the Red River for the Rio Grande.

Through the early years of America's exploration, the search for the truth was hindered by the attempts of Old World nations to keep their discoveries secret. The Europeans came as one another's rivals, bringing their sharp-edged, three-cornered hatreds —national, religious, and racial. They extended their bitter competition for power to their empires overseas. Knowledge about new lands, possible sources of wealth, and trade routes were closely guarded pieces of confidential information.

The Spaniards were among the boldest explorers. But they passed on little of their findings about America to the world at large. Many of the original maps they drew were never copied or published. Instead they were kept in locked vaults at Seville. It was an old custom on Spanish ships to weight maps and charts with lead so that they could be quickly sunk to the bottom of the sea in case an enemy should try to board. The safest policy was to put nothing in writing.

The English came as colony builders, the French as fur traders.

A map of the New World printed in Switzerland, 1540

But Spain sent its treasure seekers. The Spanish discoverer Juan Ponce de Leon scoured the beaches and the swamps of Florida in search of a legendary "fountain of youth." Dozens of other parties found not the gold they sought but death in the wilds along the Gulf of Mexico.

Spaniards mapped the Gulf Coast and said it "bends like a crossbow"—with the Mississippi River as its arrow, pointed toward the North Star. There were times when Spanish and French parties wandered near each other through the piney forests along the lower Mississippi, neither knowing of the other's presence.

Many of America's early explorers were bands of men who crossed the land plumed, armored, and mounted. Some passed through like a whirlwind, others like a shadow. Caring nothing for the countryside, they neither mapped nor sketched nor surveyed. They drove through the wilderness with set jaws, their eyes eager for the glint of gold. Occasionally they paused to take possession of a region in the name of some imperial majesty. The wilderness echoed these same claims again and again in different languages, seizing ownership of the same piece of land.

The banners of European monarchs floated over vast regions. Often there were no limits to the territory claimed. No maps were drawn. The boundaries were unmarked. No acceptable agreement or treaty was made with the Indians who actually lived on the land. Sometimes these uncharted properties were turned over to new owners who knew even less about their exact location. No wonder America became a bloody battleground of ancient ill will!

In the end, however, the violence and conquest counted for little. No foreign power was able to maintain its rule. The country was won finally by those who made it their own by taming the wilderness, staking out homes, replacing nonsense with knowledge, learning the truth about the land and what it had to offer.

There is, of course, no one map of America. What we have is a great treasury of mapmaking, the work of countless people, going back hundreds of years in time. The mapping of this land began with the earliest residents who saw some need and some way of making a record of what they had discovered. From the first crude attempts to the work of today's highly skilled mapmakers, this has been a continuing effort. Each in his turn added and improved, expanded and enlarged on the mapmaker who went before him.

Men had arrived on this continent long before Columbus. Almost all their maps have vanished in the mists of time. Diggings in the earth, however, reveal the traces of mapmaking skills that go back a great many generations.

Among the fifty states of the United States, the last two which achieved statehood were the first two regions to be explored and settled.

It was more than two hundred centuries ago when men first journeyed from Asia into North America. The Bering Sea now filters through the series of steppingstone islands between the two continents. But once there was a solid land bridge on which men crossed into Alaska, probably in pursuit of game.

About ten centuries ago, South Sea islanders in small craft crossed the Pacific to populate the Hawaiian Islands. Some time later Viking explorer Leif Ericsson landed somewhere on the Atlantic Coast of North America. He stayed briefly and never returned.

From these early discoverers we have no existing maps of America—except for the tattered remnant of a map that the Vikings drew, showing what they called Vinland.

The early arrivals on the Pacific side seemed to use maps in special ways. They were probably less interested in mapping the entire world than they were in solving practical problems of navigation, food gathering, and home finding.

The South Sea islanders developed methods of mapping that answered the need of how to find one's way across the open sea. Their maps, woven from twigs and shells, showed the stars and the ocean currents which could be used as guides in moving from one island to another.

The Alaskan Eskimos were capable mapmakers, with skills going back to ancient times. In charting their regions they used many methods to show the shape of the seacoast, the overland portages between the waterways, the distances in terms of days of travel. Finely carved pieces of bone and stone have long been used to describe the contour of the Alaskan shoreline.

The American Indians had well-developed mapmaking techniques long before the arrival of the white man. There are still in existence old maps on birch bark and deerskin which guided the Indians in their migrations to new hunting grounds.

From Europe, white men brought here a different mapmaking tradition. Out of the earliest civilizations of the Old World came the idea of the earth as a globe. The ancients divided the circle into 360 degrees, each degree into 60 minutes and each minute into 60 seconds. Mapping was based on a simple idea—that it was possible to locate any point on earth as the crossing of two lines.

The early Greeks provided that system of lines. They invented the grid network of parallels and meridians. Latitude, they said, is distance in degrees north and south of the equator. Longitude is distance east and west from a zero line—the prime meridian running from one pole to the other.

Mapmaking arts flourished around the rim of the Mediterranean Sea. Men learned to make accurate charts by which to navigate. In time they were encouraged to venture out on the great oceans, finding new continents of which they had never dreamed.

By the sixteenth century America was a center of world-wide curiosity. An age of bold seafaring and exploration, this was also an age of mapmaking, when every decade brought the New World into clearer view. Each new map of America sent a wave of excitement quivering across Europe.

A Bavarian metalworker named Gutenberg perfected the molding of type and thereby unloosed a flood of printed matter. The production of maps remained a slow, painstaking process. Working from a drawing, craftsmen tediously cut each fine line into a metal plate or chiseled a wooden block. However, the new art of printing made numerous inexpensive copies of maps available. It was in this period that maps came out of the royal palaces, the military headquarters, and the libraries of scholars and into the hands of the people. Many persons learned to read maps. They came to understand that a map is a framework of knowledge about the the earth's surface or a portion of it. A map, they realized, offers a graphic report about some place in this world, its location, its size and shape, and the form and features of it.

A map is a picture which draws on the imagination. Map reading is the skill of looking at lines, shadings, colors, symbols, and seeing there a vision of the landscape as it really is. The word "river" on this printed page is more than just an arrangement of black lines on white paper. Somehow that printed word conveys to us something long and flowing and natural and wet. The same kind of idea is expressed by the line on a map which symbolizes a river.

With a little imagination it is possible to look at a certain kind of map and see the land depicted as though it were a view from high above. In the mind's eye appear the uplands and streams, the towns and roadways, the shore lines and forests. Cross-hatched lines are soaring peaks and blue threads are tumbling white-water streams.

The inner reality of the map is entered by means of a "key" which the mapmaker provides. Also called the title box or legend, this guide leads to a deeper understanding of the map. Here may be found the graphic scale which says, in effect: "One inch on this map represents this many inches or miles on the earth's surface."

The legend may include many other helpful facts, such as the meaning of the colors and the symbols used. Directions are also indicated, although the usual practice is to have the north arrow pointing toward the top of the map. In the legend the mapmaker often tells the kind of framework or projection which he used in order to picture the rounded earth on a flat surface.

Even the earliest maps of America, with all their faults, served to make the New World understandable. The maps helped to turn the shapeless wilderness into something which men could visualize.

In spite of their meager information the European mapmakers produced a remarkable series of attempts at picturing the newly discovered regions. True, there were gaps where nothing appeared. Areas where the land was completely unknown were marked *terra incognita*. At times the mappers invented features in the place of missing knowledge. Or else they filled the empty spaces with decorative scrolls, pictures of unreal-looking Indians, fabled beasts, unlikely sea monsters, and puffy-cheeked wind spirits. Few of the important European mapmakers ever crossed the ocean or set foot on American soil. They had to rely on scattered bits of information, on secondhand reports, including hearsay, rumor, and half-truths.

At the same time a new breed of mapmaker was appearing in the American wilderness. He mapped what he saw, then pushed on deeper into the country to set down new truths. He lived on the land, sailed its streams, climbed through the folded patterns of the uplands, studied the sweep of the plains—and drew his own picture of it all. When he had passed through and mapped it, the land sprang into clear view.

Many of these men began life as soldiers, trappers, adventurers, frontiersmen. Few were trained in mapping, nor were they learned in science or mathematics. But somehow they were struck with the wonder of this new country and took delight in recording their discoveries.

They were challenged by myth, mystery, and misunderstanding. In their mapping, their purpose was to tell something about America which could not be told in any better way. As they came to know portions of this country firsthand, they copied directly from nature. Bit by bit they built up a body of truthful geographical knowledge about America and about the United States as it took the form of a new nation.

The greatest of our mapmakers added long lines of discovery and large areas of exploration to the map as it slowly took shape. As they worked they lived some of America's most exciting adventures. This book is their story.

2.
The Mapmaker on the Rock

For a hundred years after Columbus, few European families dared to settle on this continent. It was not that they lacked courage or a spirit of adventure, or were unwilling to leave their homes in Europe. Many were, in fact, all too ready to escape the wars and religious feuds in which Europe was embroiled.

But the New World had yet to be explored—and mapped. The mapmakers had to set down the pattern of the landscape and the contour of the shores, the rise of the ridges and the course of the rivers. Until then Europeans knew little about this continent—much less than we now know about the moon.

It is surprising, perhaps, that Samuel de Champlain aroused so little suspicion in that year of 1599. He was, after all, a young Frenchman wandering about through the Spanish possessions in the New World—where he had no official business. He carried with him at all times a notebook in which he quietly jotted bits of information. Wherever he went, he mapped and sketched in detail the seacoasts and the settlements, the terrain, the harbors, the forts, everything.

In another time, under other conditions, Champlain would surely have been seized as a spy—and hanged. But according to the stormy European politics of that moment, the Spaniards were friendly with the French and hostile to the English. And Champlain was allowed to go his way, gathering up the secrets of America as he went.

The sixteenth century, just ending, had opened the Age of Discovery. This great double American continent had been newly found. The entire globe had been circled. Every year of the century yielded exciting new discoveries as the European powers vied to outdo each other on the high seas.

Champlain recalled how people in his native village had hailed Jacques Cartier, who had founded and explored New France, far to the north. But Champlain's native France was never a first-rate maritime power. Only a few Frenchmen of the sixteenth century had raised that nation's banner in the remote outposts of the earth. France had, in fact, produced no explorers to compare with the Spanish *conquistadors* and no seafarers as bold as the Portuguese Magellan or the English Sir Francis Drake.

As for Champlain himself, he knew he was not likely to find any hidden worlds. His style of work was modest and methodical. He was stronger on diligence than on daring. Yet something within him responded to the outbound spirit of his age. He had a driving need to play some part in it. And if Champlain could not discover the new lands, he could map them.

There was an enormous demand for what the French called a *carte,* the English a chart, the Spanish a *carta.* A profession of growing importance was that of chart maker. The science called cartography was centuries old. But the Age of Discovery, Champlain's own exciting age, produced a large body of men who made mapping their life's work.

The mapmakers were, in fact, making some history on their own. The sixteenth century began with a great world map by Juan de la Cosa, the cartographer who had traveled with Columbus. It showed the New World merely as a series of ocean islands, steppingstones to the Orient.

In 1507 a German mapmaker, Martin Waldseemüller, gave a name to the discovery of Columbus, calling it America after the Florentine explorer Amerigo Vespucci, who had been among the first Europeans to see the mainland in 1499–1500.

Each year this land mass grew in size on the maps which were beginning to pour out of the engraving shops of Florence and Antwerp, Nuremberg and Paris.

As a schoolboy Champlain had been impressed with the maps of the great Flemish cartographer, Gerardus Mercator, who showed two distinct continents of great breadth labeled North and South America. But to his puzzlement, the young Champlain also saw other maps which pictured America as a long, narrow strip, cut through by a number of water passages.

Growing up in the French seacoast village of Brouage, Champlain set his face toward the Atlantic horizon. Beyond the sunset were newly discovered lands of indefinite shape and form. The winds out of the west whispered to him of this New World.

His father, a naval captain, brought home to the fireside the current tales of seagoing bravado. Even the fishermen in the village talked of little but the latest discovery beyond the seas. Each new landfall seemed to fill the people with a thousand questions. The amazing stories brought back by European explorers

Finding the altitude of the sun with a cross-staff

were told at court, spread through the cities, echoed in the villages, and carried out into the countryside. They stirred the minds of men. But many such tales stretched the truth and put a strain on belief.

What made them real, however, were maps. If a man could spell out a strange place name on a map, trace its outline with his finger, reckon the distance and direction from some familiar point—only then did rumor become reality.

No previous age had ever had such a vigorous interest in the world at large. Nor had there ever been such a clamor for information about foreign lands.

Early in life Champlain learned the arts of navigation. He became skilled at finding his position on earth by astronomical

Using an astrolabe at sea

observations. He learned how to use such instruments as the cross-staff and the astrolabe for measuring the altitude of heavenly bodies above the horizon. The boy moved easily into the methods and skills of cartography. From that time on his map-making tools would forever be part of him, like his hand or his eye.

Fascinated with maps, Champlain admired the learned men whose work brought together the latest available knowledge of the earth. These cartographers designed new globes and world maps. They were scholars who had mastered mathematics, geography, and astronomy. Many were skilled in printing and engraving.

But Champlain saw a different kind of life for himself. He

could not be happy in some studio, gathering bits of information to be added to existing maps of the world. For the young Frenchman, the mapping of distant lands meant one thing. He must go there himself, see it with his own eyes, record it with his own hand. By the time he was ready to leave home Champlain's mind was made up. He would devote himself to mapping the New World.

Once he managed to get passage across the Atlantic, Champlain jumped from one Caribbean island to the next. He spent some time in Puerto Rico, mapping that island. He reached the mainland at Vera Cruz. From there he followed the dusty trail of the Spanish conquerors from the seacoast to the distant highlands of Mexico City.

Wherever he went Champlain found these lands filled with magic and mystery. In Mexico he learned of the beautiful quetzal bird. According to an old fable, it was born without feet and came to earth only in death. As he moved southward he described in his notebook the work of the pearl divers bringing up gems from the tropical waters. On his way to Panama his ship was attacked by pirates who preyed on the cargo vessels and the settlements along the Spanish Main.

A year, and a century, had come to a close. In Panama, Spanish cannon thundered in the New Year's holiday of the year 1600. Long-winded toasts were offered in praise of Spanish arms, in high hopes for Spain's future in America. Though he joined in the merrymaking, Champlain could see that the Spaniards were already in trouble in this overseas empire. The English, gaining swiftly in the rivalry on the seas, were challenging Spain everywhere, even raiding Spanish colonies in the Caribbean. Spain, the young Frenchman observed in his notebook, had made little headway in building colonies in America. Cruel treatment of the Indians had not succeeded in winning religious converts. To Champlain it seemed that the Spanish policy of the sword and the cross was a failure.

He learned of a party of Spaniards setting out for the Pacific coast. The young Frenchman was soon in their midst, hacking through the tangled rain forest of the Isthmus of Panama, scaling jagged peaks, sloshing through green scum swamps.

Rough as it was, the journey was surprisingly short. Here Champlain saw America as a strip of land between the Atlantic and the Pacific which might be easily spanned. The young mapmaker could envision the digging of a ship canal across the narrow neck of Panama which would greatly shorten travel between East and West. Such a canal, he noted in his journal, would cut the New World in half "so that the whole of America would be two islands."

Before he went back to France, Champlain had set himself the goal of returning to this fascinating continent to stay. Within a few years he was back in America, setting up a home base in the French territory near the mouth of the St. Lawrence River. From New France he explored and mapped the coast southward into what was to become New England. His hope was to find the waterway which would take him across the breadth of the continent. This was the dream he would follow the rest of his life.

It was July, 1605. An Indian fire flared wildly in the offshore wind that cooled the night. For some hours the small circle of men had talked earnestly—more with gestures than with words. The French visitor wanted to know the shape of the shore line southward and how distant were the hills against the western sky.

In the mind of Samuel de Champlain, the New World was an unfinished map. From the frontier cabin far to the north which was now his home, he roved restlessly across this wilderness, sketching as he went.

Champlain's small ship had appeared to the Massachusetts Indians like a distant gull flying low along the horizon. The place where the white men dropped anchor was later to be known as

Cape Ann, an outcrop of granite on the jagged seacoast. The Indians, encamped in a summer village for fishing, were as full of questions as the visitors. But a chasm of language stood between them and speech was a useless bridge.

Champlain snatched a charred stick from the fire. On a deer hide he made a few sweeping lines, pointing between strokes toward the seacoast. Would the Indians understand this as a map? Did they see how his drawn lines made a picture of the landscape?

After a brief moment of silence the old chief took his turn at this game. He grasped the charcoal and continued Champlain's lines. There to the south beyond the cape, he seemed to indicate, the coast makes a long arc. To Champlain's surprise, the Indian then sketched in a wide river, tracing its source far inland. The Frenchman had passed by the mouth of the Merrimack without ever noticing it.

Suddenly one of the braves raced down to the beach and brought back two heaping handfuls of small stones. As Champlain watched, the Indian carefully piled a number of the pebbles on the deerskin map. Then he showed by gestures that the mound represented his own tribe and their fishing village on the seacoast. Next he made a similar pile and still another, until there were six in all. To Champlain it was clear that this unusual map now showed the six Indian tribes which lived in this region.

Champlain was delighted by this useful exchange of information. The Indians had proved again something that he had always known—that a map can tell more than words can.

In the days that followed, Champlain toured the long arc of Massachusetts Bay, filling in the details on his maps. The light sailing ship lingered in Plymouth Harbor. Here Champlain took soundings with a lead and line, carefully noting the water depth at forty places in the harbor. He sketched the river inlets, the islands, the long and knobby headland which sheltered the bay. The mapmaker went ashore to take readings with his astrolabe and compass. And as he wandered through the nearby woods and across the beaches, he sampled the walnuts, the currants, and the wild green grapes.

Here too the Indians were friendly and helpful. And Champlain drew in on his map the places where Indian families lived and planted their corn. These Indians knew every detail of this region. But they had never found it necessary to develop the skills for mapping any wide range of it. They were travelers, but the areas through which they seasonally moved were limited. And they carried the map picture of the region in their heads, learned through long familiarity.

Through their reports of white men, some of these seacoast Indians had already come to distrust them. From such tribes Champlain's party met with suspicion and violence. More than once the group clambered back aboard their ship just in time to escape with their lives. But most of the Indians Champlain met that summer responded helpfully to this man. They sensed that his whole purpose was bound up in learning as much as possible about the land which they loved and setting down its image with as much accuracy as he could.

Summer was ending, and cold winds warned the journeying Frenchman that it was time for them to turn back toward their home in New France. But Champlain urged the party farther southward. He found Cape Cod in the shape of an upraised and flexed arm, jutting far out to sea. The expedition probed the crook of the arm, sailed around its fist and down almost to the bent elbow, then doubled back.

All the way Champlain mapped and sketched. Ashore he felt at home. He wandered with his sketchbook over dunes and sand flats, salty bogs and marshes. These lonely beaches recalled to him the French seacoast where he had spent his childhood.

During one night great thunderheads had rolled up in an angry sky. That morning Champlain came up from the ship's cabin, only to return for his greatcoat. An icy wind was blowing out of the northeast, and the sky was black. For the next four stormy days, the little ship rocked perilously in the churning white water of Cape Cod Bay.

Slowly the party made its way northward, hugging the coast but wary of its rocks. When the sun reappeared Champlain was able to go ashore at a number of places along what is now the Maine coast. Here he was fascinated by the well-developed farming of the Indians. Champlain stopped to jot down in his notebook how the Indians planted their corn in little mounds of earth, mixing the seeds with bean seeds. "When they grow up," he wrote, "the beans interlace with the corn, which reaches a height of from five to six feet, and they keep it very free from weeds."

Champlain's main interest was in charting the coastline, but he had an eye also for the natural beauty of the region and for the Indian way of life. He hunted and fished in clear streams. He was struck with the majesty of the widely spaced oak trees and noted that "one would think they had been planted by design."

On his map sketches Champlain recorded an ascending series of offshore islands which seemed to come marching up out of the sea to form a rising range of hills on the land. He was fascinated by the force of the raging surf, by the endless war between tide and rock. What Champlain mapped on his journey was a wild, wave-battered shore line with a few sheltered bays. No white man yet lived anywhere along its entire length.

But where Champlain made a remarkably accurate map of Plymouth Harbor, a party of Pilgrims would arrive within fifteen years. They would settle here in a blustery midwinter, fretful over the storms that blocked their hope for a home further to the south.

A lighthouse at Cape Ann would someday guide the Glouces-

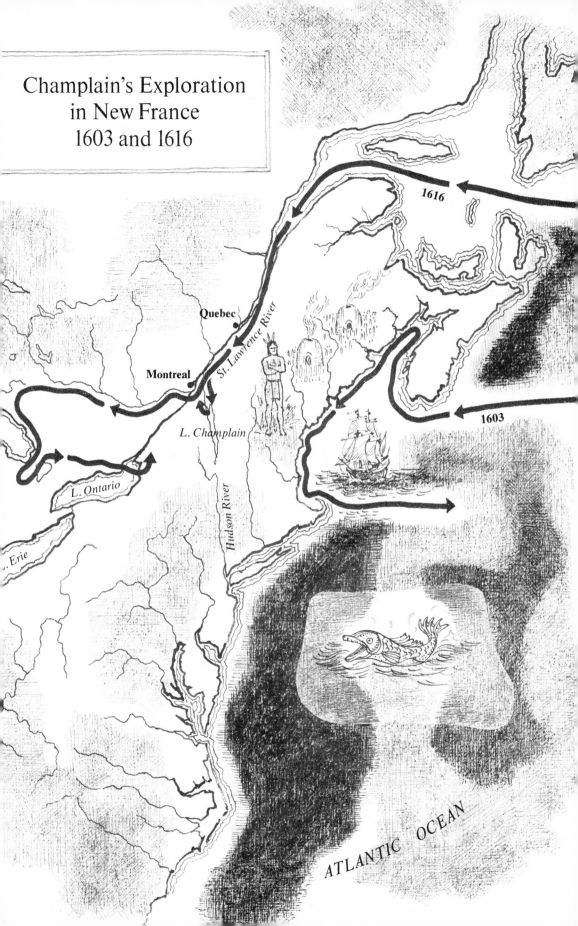

Champlain's Exploration
in New France
1603 and 1616

1616

Quebec

Montreal

St. Lawrence River

L. Champlain

L. Ontario

Hudson River

L. Erie

1603

ATLANTIC OCEAN

ter fishermen along the perilous path through foggy shoals to the Grand Banks. Provincetown, at the tip of Cape Cod, would be the home of deepwater whalers.

Where Champlain took soundings in Boston Bay, a city of rebellious men would rise to ring out the cry for liberty. In protest against the British tea tax, a band of Bostonians would dump British cargo overboard, turning these same waters into tea.

This barren seacoast would someday be peopled with flourishing cities. But now wild turkeys drummed in the woods where Portland and Portsmouth would arise, and southward-flying geese honked over the wilderness which would become populous New England.

But settlement of this continent had to wait for its exploration. And the settlers would not begin arriving in any numbers until they had seen the land laid out on a map they could trust.

For the next full year Champlain worked on his map. Any visitor to his small cabin in New France would have had to step carefully between the sheaves of papers. Laid out around the mapmaker were his small sketches. The Atlantic shore line was depicted here in bits and pieces. Reefs and shoals, bays and inlets, capes and coves—these were set down in his drawings.

And now, with patience and painstaking effort, the parts of the puzzle were to be fitted together. But where was the pattern that would guide the mapmaker to the proper order of his material?

From earliest times mapmakers have overlaid the earth's surface with a system of guidelines. The east-west lines show latitude, or distance from the equator. Each line marks off one degree of latitude, beginning from the equator. And ninety degrees of latitude covers the distance from the equator to the North Pole. Another set of imaginary guidelines was pictured as running from pole to pole. These were the meridians, used to measure longitude in degrees as distance east and west of a zero meridian.

Champlain had keyed his sketches to this grid system by setting down the latitude of every location where he made a drawing. Each sketch was a record of some place where he had stood and carefully traced the scene as far as his eye could see. He had charted the rise and fall of the terrain, estimated the distances and directions to peaks and headlands, taken readings with his instruments on the sun and the stars.

His astrolabe had showed the Plymouth harbor to be located at 42 degrees north latitude. Going northward, he had taken a reading of 44 degrees at the mouth of Penobscot Bay, where he went ashore to follow Indian guides through a land of whitewater streams and crystal lakes. At 49 degrees north latitude, he had re-entered the St. Lawrence River on the last leg of the journey to his home in Quebec.

His numerous latitude readings made it possible for Champlain to assemble his sketches into a continuous chart of the shore line. When he was through he had a large map showing a thousand miles of the mid-continent's coastal region.

On August 11, 1607, Champlain boarded a ship headed back to France. Under his arm was a roll of parchment, his map of the American seaboard. In Europe it would be unfurled like a banner proclaiming a new realm and a new age.

From Old France, Champlain returned the following year to a new home he built on a site in New France which was clearly a mapmaker's choice. He selected an immense rock overhanging the St. Lawrence River.

From its heights Champlain had an aerial view of the vast region. The great river below surged toward the open sea, giving him access to the whole eastern shore of the continent. Upstream, the St. Lawrence was to take this French pathfinder deep into the heart of America.

The rock was Champlain's *Quebec*, named with an Algonquin

Indian word which means "the place where you go back." Here his men built a tiny fort and set up a small colony which was to become the capital of New France.

The winters were bitter cold and tediously long. Fierce winds howled down out of the highlands to the north. For weeks at a time the entire region was snow-blocked and iced in, silent and motionless. In these periods Champlain worked hard at his maps, staying close to the log fire, bundled in fur robes.

By today's standards Champlain's methods, calculations, and instruments were crude. But he used them effectively to produce a series of maps which are considered remarkably accurate for their time.

As that winter dragged on into March, Champlain paced his small cabin, cramped and eager to set out again. On the twenty-first day of the month Champlain took careful readings on a dim, cold sun.

The day has a special meaning for astronomers, navigators, and mapmakers. For this is the equinox, when the sun at high noon is seen directly over the equator. Wherever he is in the world, the observer needs only to measure the angle between the noon sun and a point directly over his head. That angle is equal to his latitude.

The tireless explorer had already turned a great area of the North American wilderness into charted territory. But he waited impatiently as the spring thaw turned the river below into a wild torrent. Great ice floes tumbled headlong toward the sea. Soon, Champlain knew, this same river would be his pathway westward into lands no white man had ever seen.

That spring of 1609, Champlain began a series of explorations into what is now upper New York state and the eastern Great Lakes area. Alternating river travel and overland portages, Champlain and his Indian companions moved swiftly southward.

The landscape changed as the explorer pushed deeper into the interior of the continent. Broad-leafed trees became more numerous among the tall, pointed pines. The area abounded with cold, deep lakes.

Long after dark the party moved on through one summer night. Suddenly they came to a bright expanse of moonlit water. This large lake, where Champlain drowsed that night to the hum of insects and the soft sound of the wind, was later to bear his name.

On this expedition, Champlain became deeply involved in the wars raging between the Huron and Algonquin Indians along the St. Lawrence and the Iroquois nations to the south. The Frenchman had little choice in whether he should or should not become part of the Indian rivalries. Champlain needed the Indians to help him reach his exploring goals. In order to win their aid he had to pledge them his help and loyalty, and swear hostility to their enemies.

The Indians found in Champlain a white man who kept his word. They sensed that he was somehow a part of their world, willing to share their simple style of life. He was a kindly but tough-fibered man, shouldering heavy loads on the portages and taking his turn with the canoe paddle.

The summer ended all too quickly for Champlain. He knew that in the autumn he must return again to France. Reports had to be made to the King. And for Champlain the trip back to Europe was an opportunity to learn what others were doing in mapping the New World.

In Paris, Champlain hurried about from the ministries to the royal palace and then to the engraving shops where his maps and books were being printed. He was, by this time, a respected expert on the New World. His maps had authentic qualities not found in the works of cartographers who had never set foot outside the European continent.

Few men of his time had Champlain's grasp of the geography of North America, north and south. He had personally seen the eastern seaboard from the Gulf of Mexico to the St. Lawrence. And the continent had begun to take a definite shape in his mind.

However, as he made his way about Paris, Champlain realized sadly that the statesmen and merchants and military men to whom he spoke had far different kinds of interests in New France from his own. By the time he was ready to make his way home to Quebec he was pledged to a number of tasks. His official job was that of Geographer Royal, with the duty of making maps. But he was also commissioned to build colonies, to assist the missionaries, to help open the fur trade, and to make military alliances with those Indian nations which would help French expansion on the continent.

Champlain had no mind for business and no zeal for conquest. But he realized that in order to get support for his mapmaking and explorations, he must attach himself to those enterprises which attracted men of wealth and influence in Paris. Exploring expeditions to America were expensive undertakings. Those who paid the bills were anxious for profitable returns.

He no longer felt at home in Old France. He was eager to return to the great outreach of dark, silent wilderness into which he could bring the light of his own explorations and mapmaking.

In the years that followed, Champlain became a high official of New France. But he continued his summertime expeditions into the countryside after the harsh winters in Quebec. He was growing old now, but the uncharted wilderness seemed to call to him, and his blood stirred at each new report of possible openings to the west.

Never for a moment did Champlain give up the old hope of

breaking through the continent toward the ocean beyond, with the Orient on its western shore. No one knew how distant the sea might be. But every newly discovered river coming out of the west seemed to promise that the secrets of the Orient lay just beyond its source.

Upstream along the St. Lawrence there was a turbulent stretch where even the Indians declined to match their canoeing skills against the fury of the rapids. The water swirled foaming and white down a flight of rocky steps. Near this portage the city of Montreal would someday rise. But in the days of New France this area became known as *La Chine*, the French word for China. Men dreamed that this place would someday become a jumping-off point to the west for the Orient trade.

One summer a French youth came down the Lachine Rapids with a story of having seen the ocean. Champlain packed his traveling gear for what proved to be a heartbreaking chase after a baseless rumor.

(It was on this journey to Lake Huron that Champlain lost his precious copper astrolabe, which was found two and a half centuries later by a schoolboy. While the boy trudged along a river trail, perhaps wishing that his studies of American history were not quite so dull, suddenly history came alive. A bit of metal caught the sun, and the boy stopped and picked up Champlain's astrolabe where the great explorer had dropped it.)

By 1634 the toll of years of hardship had bound Champlain to his rock. But he never stopped seeking out men who would carry on his work. There came a day when a priest named Jean Nicolet, who had been brought to New France by Champlain in 1618, drew up his canoe at the Quebec landing and climbed the winding stairs to the high stockade. He was a rugged man, hardened in the wilderness. His eyes shone with the vision of discovery. Indian friends had told him where and how he could find "the great water." As Champlain listened, his aged heart responded to this new hope of finding a passage to the Pacific. Was it possible that this at last could be the long-sought route? And would he now be able to map the way across America to the Orient?

Champlain thrust aside his lingering doubts. The two men went to the maps and discussed Nicolet's coming voyage. There was some excited talk about what to expect on reaching the fabled Far East, the source of spices and silk and gunpowder. Champlain offered a few ideas on the subject. As the priest departed upstream, the ailing Champlain waved him Godspeed from the top of his rock.

Weeks of hard travel took Nicolet and his Indian guides far to the west. At last they came to what was indeed a "great water." Didn't Nicolet realize that he was crossing a body of fresh water? Apparently the explorer would let nothing dampen his hope of reaching the fabled empires beyond the salt seas. Days of hard paddling across heavy swells brought them within sight of land. Nicolet made a few hasty changes of clothing and dug into his baggage for a pair of pistols. The priest leaped from the canoe to greet the residents of China. He was garbed now in a robe of Chinese silk, and in each hand he carried a pistol loaded with gunpowder.

But alas, the natives who met Nicolet that day were not yellow men but red men. Nicolet's "ocean" proved to be a lake which was later to be named Michigan, from the Chippewa Indian word for "great water." And the "China" shore was to be known as Green Bay, Wisconsin.

Back in Quebec, Champlain heard the news with dismay. He realized, however, that Nicolet had made an important addition to the map, being the first white man to view Lake Michigan.

To his end Champlain lived with the vision of the Pacific

Ocean, which he had once seen in Panama where the two oceans are only forty miles apart. After that experience the explorer was never quite willing to believe the enormous breadth of the American continent.

Champlain spent his last days dreaming that men would someday find a fresh-water path between the salt-water seas.

3.
Man in a Canoe

Lay an open right hand, palm up, on the map of North America. The thumb crooks downward, following the Florida peninsula. The fingers extend northward, showing the bony skeleton of the continent's mountain ranges. The palm forms a basin. And through half-closed eyes you may even imagine the lines of the major rivers of America.

This is a simple way of thinking about this land—too simple, perhaps. But it took centuries for explorers to learn the basic shape of the continent and for cartographers to set it down on maps.

Just three hundred years ago a man in a birchbark canoe made the first great map of the central valley of America. He was Louis Jolliet.

The lives of men who map the earth are bounded by sun lines and sun dates. Two such dates are the equinoxes, one occurring in March, the other in September.

On September 21, 1645, as the sun was making the second of its two yearly appearances directly over the equator, a boy was being baptized in a roughhewn Quebec chapel. If the time of Louis Jolliet's birth was signaled in the skies, the place also has special meaning. He was to become another of the great explorers who mapped the American continent—but he was the first one born on its soil.

Son of a frontier wagonmaker, Jolliet was an orphan by the age of six. His boyhood was spent learning to depend on himself at the edge of the wilderness.

Two paths lay before him. One led to the seminary, where his studies might someday make him a priest or an organist playing the music which he loved. The other path, which he eventually chose, took him into the nearby woods, a wonderland of moose and bear, of boundless fir forests sheltering clear lakes and hiding a thousand mysteries. At every chance he plunged deep into this wild country. Often he went in the company of fur trappers. Sometimes he joined groups of wandering Indians.

Under such teaching he became an expert woodsman and waterman. He learned the fine points of survival, the Indian way of reading directions in the stars and reckoning distances in days and half days of travel. His lengthening trips deepened his love for the outdoors. But he was still a student at the seminary. And his music was much a part of him. During nights in the wilderness he seemed to hear the great organ hymns above the natural sounds of the woodland, the cries of hawks, and the chorus of winds. As he traveled his music came back to him, accompanied by the rhythmic dip and pull of the canoe paddle.

Back home in his village of Quebec, Jolliet was a stalwart figure, sun-bronzed and muscled by his weeks in the outdoors. A town official looked at him one day and made a remark that seemed to describe the youth aptly: "In a land where everything is to be feared, Louis is afraid of nothing."

On a hunting trip far to the west he learned how to build a canoe. Imitating his Indian friends, he shaped the supports from tough tree roots. Sheets of birch bark were stretched over the framework and sewn together with hide strips. Finally he sealed the seams with spruce gum and bear grease. Such a canoe was sturdy but light enough to be carried long distances. The Indians taught Jolliet how to handle it through the turbulent streams.

Without wheels or horses, the Indians of the St. Lawrence region used the waterways as a chief means of travel. Jolliet learned from them how to make his way through the north country. Gradually he found his way closer to the edge of the immense basin which is the concave center of the North American continent.

Thousands of years before, massive glaciers had moved down across this region, scooping out hollows, depositing rock and soil. When the ice sheets retreated, the melt formed the five Great Lakes and countless smaller lakes.

From the wandering Indians Jolliet learned of the many rivers that laced the land. He heard the Chippewas speak of the *Mee-zee-see-bee*, the river which was "the father of waters." The more he heard about that great river, the more he realized that it was probably the passageway into the inner continent.

As a French subject Jolliet felt pride in the way the French had boldly made their way into the interior of North America.

The Spaniards to the south had made little progress north of the Gulf Coast and the Southwest. The English had locked themselves into their seaboard colonies, hardly interested in moving beyond the Appalachian mountains. By contrast, the French had swept westward to the Great Lakes. A vast, rich country stretched out before them, shared only by sparsely settled Indian tribes. Outcroppings of copper and iron indicated deep mineral veins to be dug out. The wilderness abounded with enough bear and beaver to clothe all of Europe in furs.

However, no one had yet charted the way into the midlands of America. Its vast reaches remained locked in darkness and mystery. The imagination of young Louis Jolliet was stirred by the Indian talk of the great river. If such a river existed it might be the avenue for exploring and mapping the heart of America.

By his twenty-first birthday Louis Jolliet had set a few things firmly in his mind. He would someday test his exploring skills and his canoe against the challenge of the Mississippi. But first he must learn something about the art of cartography so that his voyage would produce not merely an exciting personal adventure but a map of value to New France.

Back home in Quebec there were friends who responded helpfully to his new plans. It was late August when the young man boarded a ship. He was dressed in borrowed clothes and he had borrowed money in his pocket. But he was on his way to France.

In Paris, Jolliet's eyes were opened to many things. He was disappointed to see how poorly the Europeans understood the geography of America. Somehow everyone seemed to remember the old story of Balboa, the Spanish explorer who had scaled a high peak in Panama where he could view both the Atlantic and Pacific Oceans at once. In France many still looked on America as simply a narrow land barrier between the two oceans.

Jolliet was questioned about his idea of exploring the length of the Mississippi. Was it possible that the river drained into the Pacific Ocean? Or perhaps the Gulf of California? Jolliet didn't know the answer. But he was more determined than ever to find it.

That year in Paris gave the young man some schooling in cartography. Paris was at that time a center of the mapmaker's art. A great new map of America had just been produced by the noted cartographers of the Sanson family, a father and two sons. Jolliet was interested to see that although their map showed the five Great Lakes, there was no trace of the Mississippi River, as he had heard it described.

Back in America, Jolliet spent the next few years in exploratory trips, supporting himself by joining in the fur trade. He improved his knowledge of Indian languages and made friendships among the members of many Indian tribes. Jolliet could now better understand the attitude of the Indians toward the land and their growing fear of the encroaching white men. To the Indians the earth was a gift of the gods, to be shared by all mankind. While they believed that everyone was entitled to gather the bounty of the hunt and the harvest, the red men did not understand anyone's claiming the right to own the land itself.

On a sun-warmed June day in 1671, his wanderings brought Jolliet to Sault Ste. Marie. There a mission had been built at the point where the three largest of the Great Lakes came together like the petals of the French fleur-de-lis. Jolliet came out of the woods with some Indian friends to attend a brief ceremony. In the center of a circle of white men and Indians was a representative of Louis XIV, King of France, garbed in velvet breeches and a satin sash. François Daumont had arrived in this wilderness to take possession of it in the King's name. A sword was drawn, a cross upraised. A prayer was intoned.

Daumont claimed for France all the Great Lakes "and all other countries, rivers and lakes, and adjacent tributaries, those discovered as well as those to be discovered which are bounded by the two oceans. . . ." After the King's royal coat of arms was nailed to a tall cedar post, the crowd disbanded. Each had his own private thoughts about what he had witnessed.

For Jolliet there must have been some doubts as to whether there was really any meaning to such an occasion. He had lived his far-ranging life as free as a deer. He was a wayfarer across the lands and the streams who gave little thought as to who owned them. If men were to thrive in this vast land they must explore it and map it, settle there and become a living part of it.

Jolliet could only guess at the reaction of his Indian friends whose ancestral homelands had just been "claimed" in the name of a foreign monarch who had never set foot on any of it. But he was hardly surprised at the amusing news he heard the following morning. Sometime during the night, the King's royal coat of arms had disappeared from the post.

More than a full four seasons passed before Louis Jolliet returned to the gathering place of America's greatest lakes. It was December now, and Jolliet was a solitary figure, heading his canoe into the gale-driven sleet. At his back was Lake Huron, and ahead, just beyond the Straits of Mackinac, lay Lake Michigan, the starting point for what was to be the great event of his life.

At the moment, however, Jolliet was straining for the sight of a tiny mission somewhere on a lonely headland. He was caught in the eye of a whirlwind. A violent storm raged about the low land bridge between the three Great Lakes, and the blizzard wailed down across Lake Superior.

Jolliet wondered now whether he would find his friend still waiting for him so long after the season for travel. On the difficult journey from the East, his time had run out. But he had pressed on into the winter weeks, drawn by the thought of the adventure which lay ahead.

At last there was a light. Through the storm the young explorer could dimly see the chapel, almost buried in snow. Father Marquette was startled to have his friend suddenly appear out of the wild, black tempest. But when the traveler emerged fully from his blanket coat and fur hood, there was Jolliet's flashing smile and his breathless cry of greeting.

Jolliet was still in his twenties, ruddy and dark-eyed. He was sturdy as a northern spruce and as rugged as the north country itself.

Although months would pass before these two men would leave on their long journey, Jolliet lost no time telling the priest

of his plans. They would leave with the first spring thaw, crossing Lake Michigan to a wide inlet which would someday be known as Green Bay. From that landing on, the rest was unmapped mystery. Indian tales reported waterways to the south and west. There would be overland treks and fast-running streams to ride. But somewhere beyond was the Mississippi, the broad river whose upper mainstream had never been seen by a white man.

Where that great river flowed was unknown to Jolliet. Did it empty itself into ocean or gulf or undiscovered lake? What was its course? What dangers lay along its banks or threatened voyagers in frail canoes? All these things were unknown.

But the young explorer and the slender, black-robed priest took a vow that night. With their handclasp they pledged to bring home a map tracing the flow of the "father of waters" and the valley which it drained.

As the winter months dragged on, the expedition took form. Additional men were recruited for the journey, five seasoned frontiersmen besides Marquette and Jolliet. Two well-made birch canoes were prepared. Rations of Indian corn and smoked meat were stored up. The priest readied his religious articles, including a small traveling altar which he could strap to his back. Jolliet gathered together his mapmaking supplies—compass and astrolabe, dividers and drawing equipment.

Gradually the late spring made its appearance. The morning sky darkened with wildfowl moving northward. Trout headed up the small icy streams to spawn. The woodland smelled of buds and blossoms. Along the horizon Jolliet saw the first stars of the summer constellations. By the first of May, Jolliet had discarded every item of equipment the party could possibly do without. And by mid-May they were waving good-by to the last wilderness outpost of New France.

On the western shore of northern Lake Michigan the seven men in two canoes paused. That evening Jolliet hung his astrolabe on a tree branch and took a reading of his position. They were at 45° north latitude, exactly midway between the equator and the North Pole.

Jolliet noted that fact on his map as he sketched the pattern of the landscape. By now Jolliet was an experienced mapmaker. However, mapping in the field is no simple undertaking. Things appear to be different from the way they are. An island may look as though it is attached to the mainland. Distant hills seem closer than they really are. A river may unexpectedly change direction so that it appears to meet itself coming and going.

They passed now from the Great Lakes to a land of small lakes and swift, cascading rivers. The two canoes moved through a richly patterned landscape strewn with great boulders. Deep ridges and valleys marked the edge of an ancient ice sheet. Suddenly the river itself vanished underground, leaving the voyagers stranded amid sheer purple cliffs.

Jolliet called a brief rest. But soon the men shouldered the canoes and trudged along an old trail where generations of Indians had moved from the end of one watercourse to the beginning of the next. Someday this place would be the site of the tree-shaded city of Portage, Wisconsin.

As the June days lengthened Jolliet led his little group through a green wilderness. The river carried them steadily south and west, into the setting sun. On the afternoon of June 17, 1683, the two canoes were suddenly swept into a powerful tide. The Mississippi!

Paddles stopped as the men gazed in silence at the awesome river. Jolliet, his heart pounding in his chest, gave a wild shout of joy. It was the kind of experience of which men dream. This was the moment of discovery that rewards years of hope and weeks of hardship. The first great goal of Jolliet's life was now achieved.

Swollen by the rainy spring, the Mississippi rolled majestically southward. Lining its wide banks were tall palisades of limestone. Jolliet's party drifted downstream, filled with wonder. From the subject of Indian fables, this magnificent river had turned into a flowing reality.

There is fear in the hearts of even the most courageous of men. And Jolliet now faced soberly the real dangers to his expedition. He had seen maps which showed that a thousand miles to the south the Gulf coast was edged by Spanish missions and forts. Spain also claimed the rights to a broad river which she had named Espíritu Santo. Was this the same river as the Mississippi? And what would happen to Jolliet and his companions if they should blunder down under the guns of a Spanish fort?

Jolliet had lived as a valued friend of the Indians in his homeland. But these same friends had sternly warned him that the tribes to the south were warlike and filled with hate for white men. What chance would there be for this handful of strangers, defenseless except for a few small arms?

At their camp site on a great bend in the river, Jolliet talked earnestly to his men about the importance of the information which was being gathered on this journey. The map, he explained, would be its main result. Jolliet drew a promise from every man around the circle. If harm came to himself the map would somehow reach the authorities in Quebec.

It seemed wise now to take every precaution against surprise attack. That night, and every night of the journey thereafter, the party returned to their canoes and paddled out to midstream. In the bottoms of their craft they slept until dawn, taking their turns on watch.

While Jolliet worked at his instruments and his sketches, Father Marquette filled the pages of a notebook with his comments on the strange world around him. He had lived the scholar's life in France, and once he had hoped to be sent as a missionary to China or India. But now in this wild wonderland he experienced a joy he had not known before.

"I never saw a more beautiful country," he jotted down. "As to the forests, there is perhaps nothing in nature to compare to them." In the water below they occasionally glimpsed a huge sturgeon or a whiskered catfish. Elk waded in the shallows. Overhead passed ducks and geese. Then, as now, the Mississippi was America's greatest flyway of migrating birds.

The smell of wood smoke forewarned the voyagers of an Indian village. Jolliet cried out a friendly greeting. The white men were made welcome once the Indians had got over their astonishment at the appearance of these unexpected visitors. Jolliet's party was persuaded to stay as guests. Father Marquette used the opportunity to try to explain his religion while the red men listened politely and joined in mumbling a prayer or singing a hymn.

The Frenchmen had brought small presents, and each Indian was delighted to receive a steel knife or a hatchet. In return they presented Jolliet with a gift which was later to save his life. It was a long-stemmed pipe decorated with feathers, the bowl fashioned from polished red stone—the *calumet*, or pipe of peace. Through the ages Indian tribes weary of war had come together to discuss their differences and shared a smoke as they talked. In time the pipe itself became respected as a token of peace.

While the priest eagerly looked forward to more contact with the Indians, Jolliet was wary. The languages and the customs became more and more unfamiliar as they descended the river, the Indians less friendly. To Jolliet the main task of the expedition was exploring and mapping the countryside—and trouble was to be avoided.

The party had entered the Mississippi at 42° north latitude.

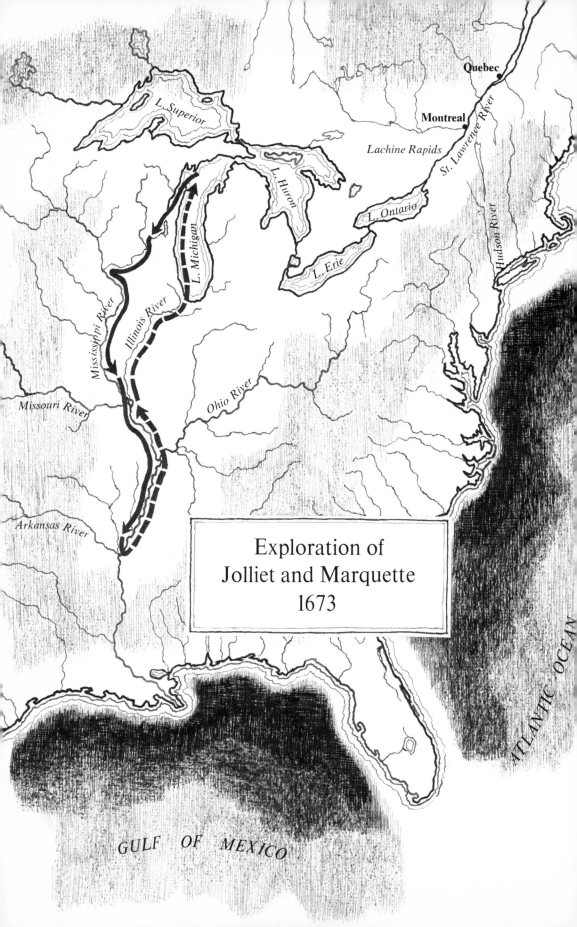

Exploration of
Jolliet and Marquette
1673

Repeatedly Jolliet measured the height of the sun and stars in the sky and checked his readings against his book of declination tables, which showed the position of the heavenly bodies for each day of the year. The degrees of latitude dropped to 41, 40, 39. And just as gradually the great river widened out, with broad wind-blown prairies on either side. As Jolliet's party passed from one parallel to the next, the scenery, the vegetation, the wild life changed.

Just below the 39th parallel the current suddenly turned violent. The river, filling with logs and branches, boiled and churned. Jolliet, in the prow of the lead canoe, was now alert for floating objects. A piece of driftwood could easily snag open the birchbark skin of the canoe and bring the Mississippi gushing in.

Using paddles to steady their craft, the canoemen fought a river which had gone wild. Downstream they heard roaring thunder like the sound of a great waterfall. It took every bit of skill and muscle now to prevent the tiny vessels from overturning. The boats were tossed from side to side, upended and pitched toward the rocky shore and then twisted out into midstream by the foaming eddies. The rapids were flinging uprooted trees with frightening force against the frail canoes.

As the waters calmed a bit, Jolliet could see the cause of the turbulence. They were passing the mouth of another great river which joined the Mississsippi—the Missouri. Jolliet shaded his eyes and followed its course westward out of sight. To the young explorer, here was another challenge. This appeared as still another unmapped corridor into the heart of the continent. But one hundred and thirty years would pass before two young Virginians would begin at this point to map the Missouri to its source.

The Mississippi, fed by numerous large and small streams, was now widening, twisting, and changing strangely. Sketching became more difficult for Jolliet as the river spread and wandered over the countryside so that he could hardly tell where its channel was and where its banks were.

Turning a horseshoe bend, they suddenly came upon an Indian village. Here large numbers of warriors lined the shore. Their clamor was far from friendly. A spear and an arrow whistled through the air. Jolliet tried to shout to them from his canoe. But he was unheard above the wild yells, the taunts, and the screams of rage. In moments the Frenchmen were surrounded by braves in dugout canoes.

Jolliet could see that they were trapped and outnumbered. Flight was no longer possible. Marquette suddenly remembered the *calumet*, the feathered peace pipe, and flourished it high in the air. The Frenchmen beached their canoes, prodded by the spears and the threats of the tribesmen.

Again and again the priest raised the *calumet*, Jolliet pointing to it and trying to calm the enraged Indians. At last some of the elders of the tribe called attention to the peace pipe and demanded that the young warriors respect it.

Jolliet and his men could feel the hostility around them as they were led to seats at a council fire. The Frenchmen and these Indians did not know the same language, but questions were somehow asked and answered. Jolliet learned that the Gulf of Mexico was ten days' journey downstream and that the Spaniards were settled at the river's mouth.

The Frenchmen were permitted to return to their canoes, but the menacing warriors still blustered around them. The braves pointed to the canoes with shrill laughter, evidently never having seen such craft before nor the bark from which they were made. What kind of animal has such a hide? they seemed to be asking derisively.

A spear was poked through the birch bark, and then another. A new outcry of anger went up. More holes were slashed into the frail craft. And now the older Indians had to intervene again. They insisted that the strangers be allowed to go their way in safety. The braves watched in sullen silence as Jolliet and his men quickly repaired the holes in their canoes. The French embarked as rapidly as they could, sighing with relief.

It was time now to reconsider plans. When they reached the inflow of the Arkansas River, Jolliet decided that the expedition had gone far enough. To risk further scrapes with hostile Indians or an encounter with the Spaniards would be foolhardy. On July 17 the party began their return trip. They had crossed the 34th parallel.

Long after the midsummer dusk, the returning voyagers camped one night under a high bluff, near the place where the river city of Memphis, Tennessee, now stands. Louis Jolliet trained the sights of his astrolabe on the North Star. He calculated his latitude as 35° north. But as he marked his position on his map, he realized sadly how inaccurate were his mapping methods.

Any point on earth can be fixed precisely at the crossing of two lines, the parallel and the meridian. But Jolliet had no means of reading his longitude in the heavens. Like most explorers of his day, Jolliet had to rely on a rough estimate of east-west distance covered from a point of known longitude. In such dead reckoning, the time spent traveling in a given direction at an estimated rate of travel was plotted on the map. If the traveler also moved from north to south, he could use the changes in observed latitude to verify his speed reckonings. Not until the next century would men accurately measure the longitude of this place as 90° west.

Throughout his history-making voyage along the Mississippi, Jolliet unknowingly remained close to the meridian of 90° west longitude. That imaginary line, running north and south through the midlands of America, is clearly shown on modern maps. But in Jolliet's time longitude was no clear and simple matter. The sun and stars were well-established guides for the measurement of latitude, north and south. But for centuries men had searched the skies for an equally useful indicator of longitude—without finding a glimmer of light.

In Jolliet's time it was generally agreed that the earth was to be measured off from east to west in meridians. But where was the starting point for such a measurement? The major European nations could not agree. Each jealously insisted on its own choice, usually its capital, as the site through which would run the meridian of zero degrees.

But even if a single prime meridian could be accepted by all, the seventeenth-century traveler, explorer, or mapmaker was faced with an even more perplexing problem. There was no way by which he could accurately determine the exact longitude of his location.

From the beginning the unsolved problem of longitude had its strange effect on the discovery, exploration, and mapping of America. Without a guide to his true east-west position in relation to the actual size of the earth, Columbus mistook the New World for the Orient. Unable to measure longitude precisely, men explored both seacoasts of America without knowing exactly how broad was the continent between them. Thus the early maps of America tended to be very inaccurate, and the wonder is that Jolliet's map came out so well.

The young mapmaker did his best—estimating the distances, noting the landmarks, sketching out the twisting course of the great river. He pinpointed the Indian villages and the mouths of streams. He noted also on his map the nature of the landscape.

The canoemen now pressed harder against their paddles as

they moved northward upstream. Near the 39th parallel Jolliet signaled his party to turn to the right into the mouth of a tributary stream. At this point they left the Mississippi and entered the Illinois River.

Through this region the place names on the map echoed the sound of the curious clucking speech of the Algonquin Indians—Keokuk, Cahokia, Kaskaskia, Kickapoo, and Kankakee. The tribes they met were friendly, often expressing amazement that this tiny party of white men would venture on so extensive a journey.

Much of the land through which they now passed hardly bore the mark of man. Along the waterways the only dams were those built by beavers. Instead of roadways, there were the worn paths of buffalo. Towns were underground, inhabited by gophers. In the skyways overhead, wild fowl winged by in great flocks.

Jolliet could foresee this valley overlaid with a patterned carpet of farms, vineyards and orchards. "Here," he wrote, "a settler would not have to spend years cutting and burning timber. On the very day of his arrival, he could put his plow into the ground."

The gently curving river, studded with wooded islands, threaded its way at times through narrow gorges. Farther upstream it widened out through flat country. Stands of trees stood like islands in the boundless sea of the tall grass prairie. Here the explorer's campfire flickered under a star-bright sky that arched 180° from horizon to horizon. At this place a city would someday be named Joliet. Its citizens would drop an "l" in the spelling and pronounce the name in un-French fashion—but there would be no mistaking their tribute to the courageous young explorer.

A series of portage trails and wandering streams brought Jolliet and his companions back into Lake Michigan. "At the place where we entered the lake is a harbor," Jolliet later reported, "very convenient for receiving vessels and sheltering them from the wind." The great city of Chicago would someday appear at this site.

The days of autumn found the explorers edging around Lake Michigan. They arrived at the mission of Father Marquette with the first snowfall. Jolliet wintered there, completing work on his map and his notes, making a copy of his great map which he left at the mission.

In the spring it was a lighthearted Jolliet who set out for his home in Quebec with two companions. One was a French frontiersman who had made the long Mississippi voyage with him. The other was a bright young Indian boy who had been sent along with Jolliet by an Algonquin chief.

From the high basin of Lake Superior, the water levels dropped lower and lower as the rivers flowed toward the ocean. Moving rapidly eastward, the trio soon crossed their last overland carry. The canoe trip down the St. Lawrence would be a swift passage. The river raced from a high plateau toward the level of the sea. Its wild course was marked with rocky shoals, treacherous currents, and turbulent cascades. But Jolliet and the burly riverman in the stern were experienced voyagers who handled their paddles with marvelous skill.

This last leg of the long journey went like the lively song on the canoemen's lips as they sped along the gushing river. Jolliet was two years gone from his home, Quebec. The young explorer was homesick. But he was also eager to report on his great voyage of discovery to the officials. It would be a message inviting settlers to the western frontiers of New France. His precious map would show the water route. There was more than enough room, he would say, for every peasant now grubbing a tiny farm in Old France.

Jolliet now began his descent of the great watery staircase, the Lachine Rapids. Here the St. Lawrence dropped swiftly by way of forty-two different waterfalls. At the bottom step was the village of Montreal where the travelers would make a brief stop to be welcomed and hailed for their achievement. The rapids were usually bypassed on foot, but Jolliet was eager for home.

In masterly style the canoemen steered their craft through the cascades, one after the other. The Indian boy gaily counted off the rapids. Ten of them, and then twenty, and thirty. The canoe darted through the swirling waters. Thirty-nine, forty, forty-one. In sight now were cabins and farms on the outskirts of Montreal.

The canoe swerved and plunged. Suddenly the three were flung high in the air and hurled into the churning stream. Jolliet reached out for the overturned craft, lost his hold. He felt the furious current hurl him downstream against jagged rocks. He was sinking, feeling all was lost, lost. . . .

Louis Jolliet awakened in a tiny cottage. Two fishermen had dragged him half-dead from the river. They bandaged his gushing wounds and nursed him back to consciousness. When Jolliet learned from them all that had happened, he buried his face in his hands. His two companions were drowned, the canoe and all its contents lost forever.

In the days that followed, Jolliet rested and recovered. But he grieved bitterly over the death of his two friends. He began now to jot down some memories of the Mississippi expedition. However, he realized that he would never fully replace the sheaves of notes and journals which he had written every day of the voyage.

As for the map, lost in the river tragedy, he now remembered the copy he had left with Father Marquette and sent a messenger to Lake Superior to bring it.

It was a dejected Louis Jolliet who meanwhile called on the high officials of New France. They had expected to see the Mississippi and its valley laid out before them on a great map. Jolliet brought them nothing. Tearfully he recounted the disaster, adding sadly, "I saved nothing but my life."

Months passed, and at last the messenger arrived from Father Marquette, empty-handed. One day, while the priest was away, the mission had burned to the ground. Nothing had been saved from the blaze.

Jolliet could do nothing now but rely on his own memories of the trip. Could he remember the places, the latitudes, the course of the river, the shape of the terrain? The mapmaker set to work.

Strangely, the journey came back to him in sharp detail. There were many errors, of course. But Jolliet's map proved to be a remarkably accurate portrait of the Mississippi. It became, in fact, the basis for the maps produced during the next decades by numerous cartographers.

When Louis Jolliet finished his map, which he entitled "New Discovery," the fabled central river of America became a reality. From copies of that map, men learned of the vast oval basin in the heart of the continent and of the river running through it like a silver thread. They could envision with Jolliet his idea of a route from the Atlantic Ocean to the Gulf of Mexico, with all the watercourses linked together by the digging of a few short canals. They could see their way into a land of opportunity, rich in game and minerals, woodland and farmland, the full bounty of nature.

Along Jolliet's water route would spring the cities of Milwaukee and Chicago, St. Louis and Memphis. The Mississippi itself would become the border of ten states. The route mapped by the young explorer would be an important pioneer way westward. In time, travel and trade would follow in the wake of Jolliet's birch canoe. The dark mystery of mid-America was illuminated by the rather crude map which Louis Jolliet drew from memory in that year of 1675.

That same year, stonemasons on the outskirts of London, England, were erecting the Royal Greenwich Observatory. There the knotty problem of longitude would in time be solved, and all the world would measure longitude from the prime meridian at Greenwich.

In 1675 another map was being prepared by John Foster, a printer in Boston, Massachusetts. It was a roughly drawn woodcut, but important because it was the first map ever drawn, printed, and published in America.

And in 1675, small groups of men were moving out across the American colonies with newfangled instruments for surveying. They had begun to measure out the land, foot by foot. And they would add many fine details to the map of America.

4.
Straight Lines
on Crooked Land

The earliest maps of America answered questions that began with "What?" and "Where?" A wild world had been discovered, and people were curious. And yet that New World remained distant, mysterious, forbidding.

The great, sprawling earth mass lay unbounded, unmeasured, unfenced—and largely unwanted. The land was there for the asking. And hardly anything was worth less than an acre of American wilderness.

In time definite land holdings appeared along the seaboard.

Pieces of property took on form and value. A wild area became part of a colony or town, an estate or a farm.

A new type of map appeared, answering these questions about the land: "Who owns it, and where are the boundaries?"

In a balmy spring dawn in the year 1683 a small striking force of Maryland farmers crept stealthily on the sleeping town of Chester, Pennsylvania. In the brief clash that followed, a few shots were fired, startling squirrels in the chestnut trees. There was cheering on both sides in the confusion over who had won the battle. This was a skirmish in one of the little-known wars of American history, a struggle over a boundary line.

In both Pennsylvania and Maryland, settlers lived on lands granted by the British crown. However, an unmarked border between them ran roughly along the 40th parallel. As quarrelsome rivalries grew more violent, the boundary seesawed back and forth. The bitterness was heightened by religious differences. Catholics and Quakers wrangled across an imaginary line.

It was another eighty years before the conflict was settled according to a map drawn by a couple of young English surveyors. In the winter of 1763–64 Charles Mason and Jeremiah Dixon arrived in the disputed territory, sent by the Royal Greenwich Observatory.

They lost no time setting up their surveying equipment and taking careful readings on the stars. For the next three years their sights were pointed due west. Across a thickly wooded wilderness they hacked a line as straight as a bullet's flight. A path nine yards wide was cleared for them as they moved steadily westward along a line of sights. Every eleven and a half miles, Mason and Dixon stopped to check their position with the stars. Cut limestone markers were erected every mile as they advanced.

Their line followed the compass bearing for two hundred forty-four miles. The two surveyors were now deep in the wilderness. But they halted only when the arrows of Seneca tribesmen began hitting uncomfortably close. The line they drew can be seen on any map which shows the southern border of Pennsylvania. That boundary was to be known in American history as the Mason-Dixon line. On one side was the North. On the other side was "Dixieland," the South. The line separated the "free states" from the "slave states" where slavery existed until the Civil War.

The Mason-Dixon Line was only one of the many man-made boundaries that began to appear on the map of interior America. Men prefer to let nature provide them with the separation lines if possible. A river or a high mountain ridge makes a clear, durable boundary. Often it helps to keep peace by leaving no doubt as to where one side ends and the other side begins.

However, such natural dividers were not always available as the American colonies grew more populous and their land needs reached out to the very edges of what was theirs. In this period of colonial history the surveyor-mapmaker was very much in demand.

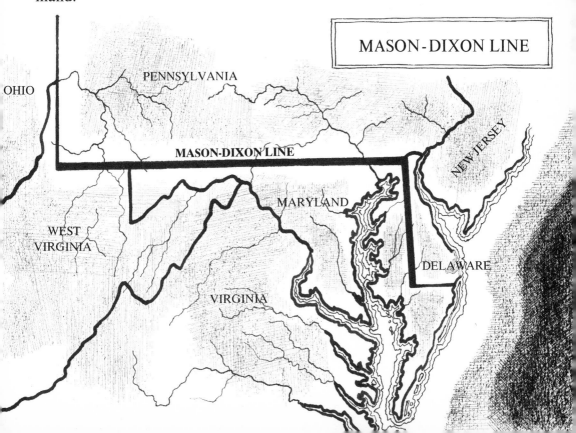

His tangled problems grew out of the haphazard fashion in which large portions of the continent were taken over. Explorers from more than a dozen European nations announced ownership rights. With a wave of the sword and a few words heard only by the birds in the trees, they claimed huge areas. The claims were often made in the name of kings who were hardly interested in the American wilderness, until they began to see benefits from it. The monarchies of England, France, and Spain used the boundless American land as a way of paying off debts and obligations. Tracts were given to friends—and often to enemies. Land grants were made to dissenting political and religious groups willing to move to the New World.

Seldom were these properties clearly staked out or mapped. The deeds were vaguely written. Often they contradicted each other, so that conflict and even bloodshed could hardly be avoided.

According to the Virginia charter this first English colony in America was given all the land between the 30th and the 40th parallels, a territory covering most of the present United States. Several additional British colonies received grants extending from "sea to sea." Massachusetts, Connecticut, the Carolinas—all had charters which placed their western borders on the Pacific Ocean.

To their dismay, the early colonists learned that an official paper with a royal seal did not necessarily give them the right to the land, only the right to fight for it. The land grants made very little impression on the Indians. America was their homeland. Many of the red men were, at first, willing to share it with the Europeans. But this attitude changed as it became clear that the newcomers had no intention of sharing. The white man sought full, outright possession.

Land ownership was an Old World idea that the Indians were unable and unwilling to understand. They had thrived on this continent for generations without the white man's idea of property and saw no reason to accept it. In New England, settlers traded with one tribe for a piece of land—only to find that they would have to "buy" it again and again from many other tribes.

The colonial years saw the beginnings of ceaseless warfare over property between white men and Indians. Land had to be taken by force and could only be held by force. Bloody conflicts ended in treaties, which were often violated by the colonists as soon as they were made.

Into this dangerous and disordered situation came the surveyor. His job was to settle land ownership problems that would not stay settled. He did succeed, however, in adding detail and accuracy to the growing map of America.

As soon as the weather broke each spring, the colonial surveyor could be found on the job with his crew. Usually he either was self-taught or had some schooling which included a bit of mathematics, geometry, and mapmaking. He owned a few instruments and pieces of equipment. Often he was a rugged outdoorsman who did not mind the withering heat of the sun or a night spent in soggy blankets on the stony earth.

The surveyor's task was to begin with a single point and lay out a line. He worked with distances and directions. The line might be long and continuous, or it might reach point after point, each time taking a new direction. Usually he completed the survey of an area by measuring out a series of lines that worked themselves back to his starting point.

The methods of survey differed widely. In the loosely clustered colonies little was done to set uniform standards. Some surveyors measured the land by pacing it off and judged angles by a rule of thumb. Often their instruments were warped and weatherworn. The markers used to show the corners and edges of a piece of land were seldom very durable. A tree, the most commonly used marker, could be splintered by a storm.

There is a typical surveyor's description of a piece of property in Vermont that reads: "Begin at the middle of a large white stump standing on the west side line of Simon Vander Cook's land and on the south side of the main road that leads to the new city. . . ."

In one boundary dispute a lawyer argued heatedly that the court might just as well accept the survey lines of a piece of property which was "bounded on the north by a bramblebush, on the south by a blue jay, on the west by a hive of bees in swarming time, and on the east by five hundred foxes with firebrands tied to their tails!"

And yet the surveyor became an important figure trudging through the American wilderness. He was a prime target for Indian arrows. Unlike earlier mapmakers, the surveyor was not able to use the watercourses as highways into the wilderness. He was chained to a line, and he went wherever the line led him. Frequently he followed his compass needle into mud-bottomed swamps, across hazardous crags, or into the nests of rattlesnakes.

But many a growing boy in colonial America dreamed of high adventure as a surveyor. He yearned perhaps to make his own mark upon the land, a mark that might last forever.

The boy, George Washington, often crept up into the family storehouse. There, amid the clutter of dusty relics, something attracted him.

He returned again and again to a number of odd-shaped black leather boxes. When he snapped open the spring locks, displayed were a set of gleaming metal instruments. These tools of the surveyor had belonged to his father, now dead. The boy peered through the slots of a sighting device and took a reading with the compass. He pictured himself leading the life of a surveyor and wondered how it might suit him.

In those years the Virginia youngster was not sure that he

would ever amount to anything. "I must do something to support myself," he wrote anxiously in his diary.

Young Washington won no prizes as a student. But in his geometry classes he was fascinated by the idea of beginning with certain known quantities and discovering a number of unknown ones. He was intrigued with the triangle. If you knew the length of one side of a triangle and the size of the angles at either end— then you could discover everything there was to know about that triangle. With a little calculation you could learn the length of the remaining two sides of the triangle and the size of the third angle without measuring them.

The schoolmaster was quick to point out the importance of this simple set of ideas in the measurement of land. It was, in fact, a key to the surveyor's art. George Washington could easily understand how a vast region might be measured off by the use of the method called triangulation. He could see himself measuring along a road, then sighting angles from the two ends of this base line toward a distant point, perhaps a church spire. From such few measurements, he could accurately map the entire triangle, and then move on to survey adjoining triangles in an endless series.

Mathematics opened a second attractive world as well. The boy studied the application of geometry to navigation at sea. He learned how to find his location by the height of the stars and to find his way by the geometry of the circle.

As his school days ended, Washington was a perplexed and serious lad. Often he wandered to the busy shipping docks at Port Royal, Virginia. His heart soared with each outbound sail. Carefully he weighed the choice between the surveyor's career and a life at sea. By his sixteenth birthday he had made up his mind.

That spring, on a raw March morning, George Washington was on his way into the Blue Ridge Mountains as part of a surveying

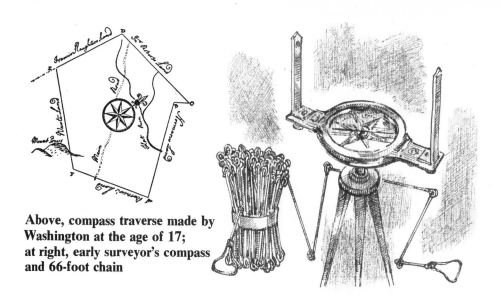

Above, compass traverse made by Washington at the age of 17; at right, early surveyor's compass and 66-foot chain

team. Gangling and peach-cheeked, the boy found himself in the company of rough and experienced frontier surveyors.

At the end of a long day's travel the party stopped at a farmhouse. At breakfast the following morning Washington's companions roared with laughter at the boy's account of his sleepless night, battling the discomforts of a straw mattress. After all, he had been raised in a house with five fireplaces, canopied feather beds, and fine china, and this was his first venture into the wilderness.

But as the days wore on young Washington began to show an impressive stamina. Surveying through rugged country, the boy soon learned the routine of the job, applying his own skill in mathematics and picking up practical pointers from his experienced companions.

One man went out ahead with a range pole. Washington operated the instrument used to sight on the pole. Once the range pole had been shifted to the desired compass bearing, the measuring was begun. Two chainmen carried a sixty-six-foot length of chain, dragging it through the underbrush between the fixed compass on the tripod and the sighting pole.

The other two members of the crew busied themselves with establishing suitable markers. For the corners, especially, the marker would have to be something durable and easily recognized, possibly a slab of granite anchored in the ground and inscribed with the necessary information. However, the boundary line itself was usually marked with blazed trees, rock piles, or whatever other landmarks came to hand.

Washington also learned on this trip the importance of the surveyor's notebook. This was a well-bound little book protected against dampness and dirt. Into it went the field notes of the surveying crew, sketches and measurements, the location of markers. It served also in the manner of a ship's log or an explorer's journal, describing the terrain, the weather, the wild life. Such a notebook became a part of the permanent record of the survey. The notes and sketches were used in order to prepare a map of the area surveyed.

During that summer of learning, the sixteen-year-old Washington also found out a good deal about himself. The Eastern seaboard was like a walled village in those days, and the boy felt himself being drawn beyond the haze-shrouded ridge of the Alleghenies. His spirits rose as the roads thinned into footpaths and the paths ended in unbroken wilderness.

As his party splashed through icy streams he discovered his own love of the outdoors and his enjoyment of exploring and mapping new country. That he could both live in the wilderness and earn his own living, he wrote in his diary, "filled me with pleasure."

Washington's party was only one of many plotting their lines of sight through the colonial regions. To the south two men were at work laying out part of the boundary between Virginia and North Carolina. They were Joshua Fry and Peter Jefferson. Partly as a result of that survey, the two would also produce the first maps of Virginia ever made. Peter Jefferson, a sturdy giant of a man, would inspire in his son Thomas a lifetime interest in maps and mapmaking.

In still another corner of Virginia a doctor named John Mitchell was enjoying his favorite hobby. His project was a map of colonial America. Dr. Mitchell put in five years of painstaking labor on this map, the only one he ever produced. It was compiled using information from all previous maps that he could find, French, Spanish, and English.

The Mitchell map of 1755 benefited from the land surveys that were taken by Washington and many others. On it the Far West still lay almost completely uncharted. The Middle West had only been roughed out in broad strokes, once Jolliet had located the great central river. But along the Atlantic Coast the sketchy outlines gave way to the close-meshed surveys.

The Mitchell map revealed how the portrait of America was becoming more finely detailed. This was the map most widely used through the following stormy decades, as the colonial armies under Washington were fighting the American Revolution to a successful end.

At the 1783 peace treaty negotiations in Paris, Benjamin Franklin would lay an important piece of paper on the conference table. It was the Mitchell map, with the boundaries of the new nation heavily outlined in red pencil.

With the last gunshots of the American Revolution, the maps of the world were suddenly out of date.

A new nation had appeared on the scene. It was, as yet, a small nation. But its boundaries were to remain unfixed and moving outward for another century. More than any other man of his time, Jefferson helped to change the map of the United States.

He was a solitary figure in that late autumn of 1783, making his way on horseback toward Annapolis, Maryland. The season's rains had turned the back country roads boggy. And as his big mare splashed along through the puddles, the statesman could feel the rawness in his chest and the weariness in his back.

He would have much preferred to be moving toward the Virginia highlands, toward home and relaxation and retirement. There at Monticello, Jefferson's handsome hilltop house was being built from his own design, and he enjoyed being one of the workmen. Also he longed for his daughters, his violin music, his map- and book-lined study where he puttered with scientific instruments, his inventions, the bones of extinct animals, his growing catalog of Indian languages. But Jefferson had saddled.up that morning and ridden off toward his duty. He was now a member of the Congress of the United States.

This wise and kindly patriot had helped to shape a loose jumble of ideas about liberty into a new type of government. Now he was turning his attention toward rearranging the handful of colonies and a vast wilderness into a substantial nation.

The map of America was much on his mind. But it was a map shadowed by darkness and disorder. For one thing, Jefferson mused, the capital of the country would soon have to settle down somewhere. In a matter of a few months the meeting place of Congress had moved from Philadelphia, Pennsylvania, to Trenton and Princeton, New Jersey, and then to Annapolis, Maryland.

Jefferson turned in the saddle and looked a long way westward. The tall, spare Virginian had been born on America's first western frontier, on the slopes of the Blue Ridge Mountains. Now he could envision an America crossed from north to south by a series of frontiers. Beyond the seaboard states and the first range of mountains was an unsettled land with the Mississippi running along its western rim. This land was part of the new nation, but it was not yet formed into states.

West of the Mississippi was an unmapped region called Louisiana, claimed by both the Spanish and the French. Its far edge was
lost somewhere in the western mountains.

Beyond the Louisiana territory was still another vast and unknown area extending to the western sea. The coast line as far
north as Alaska had been skirted by the vessels of many nations.
Most recent was the probe by Captain James Cook, the bold
British explorer of the Pacific Ocean who drew a map of the seacoast from the bridge of his ship.

At a time when few Americans could even think clearly about
the future of the small new nation, Jefferson's hopes and dreams
ranged widely from sea to sea. He would never travel westward
beyond the shrouded mountain peaks on the horizon. And yet he
would deal with each of the westering frontiers, one at a time. But
his first interest was in shaping and settling the lands just beyond
the misty Blue Ridge range. That territory was filled with
violence and discord. And the crisis grew daily as settlers poured
through gaps in the Appalachian Mountains in large numbers.

Small armies of frontiersmen were at war against the Indian
tribes, seizing their lands, setting up tiny nations of their own. In
turn the Indians war-danced in the firelit wilderness, raided the
frontier forts, pirated the Ohio River flatboats, and burned out
the homesteaders.

As far west as Illinois, companies were formed to buy large
tracts of land in schemes to set up privately owned dominions.

West of the Carolinas, mountaineers made an effort at self-
government. For a brief period their independent state of Franklin appeared on the map. Farther north, settlers formed states
which they named Watauga, Westsylvania and Transylvania. All
these attempts at statehood were in defiance of those established
states which still claimed the western lands on the basis of the
original royal land grants.

Below the surface of the young nation simmered a fierce contest. Following the Revolution, enormous power fell into the hands of the wealthy tidewater planters, the ship owners of New England, the slaveholders of the Carolinas. Battling for their own rights were America's settlers and workmen, trappers and traders. They wanted land and a voice in the government and the opportunity to enjoy the promises contained in the Declaration of Independence.

In this dispute Thomas Jefferson was wholeheartedly on the side of the men without property or power. He believed strongly that the future of democracy lay in the rough hands of the small farmers, mechanics, and tradesmen. He became their lifelong champion.

Riding into Annapolis, Jefferson found the temporary capital of the United States a battleground of many wrangling groups. The seaport town itself was the scene of a quarrel among nine states over navigation rights in the Potomac River.

When Jefferson reined up at the state house, Revolutionary War veterans were camped on the steps and in the corridors. To the quiet delegate from Virginia they voiced their anger over the government's unkept promises that they would receive free land. Feathered Indian chieftains stopped Jefferson to protest the violation of their treaty rights. Frontier leaders who had journeyed back across the mountains grabbed his lapels and showed him their petitions for statehood.

Jefferson shouldered his way through the noisy crowds. In the congressional meeting, he was plunged quickly into a series of bitter disputes over the formation of new states.

This session centered its attention on a new type of map of America. Its shore lines and uplands, rivers and towns were sketched in only the most simple manner. For this was to be a political map. Its main purpose was to show the political boundaries of the new nation and of the states within it. This was a map which had to do with liberty and how a man might be a citizen in a democratic nation. It would deal with voters and how they were to be represented. The nation, now free, needed form. And the political map would show the structure of its system of government.

To Jefferson this Congress appeared as a body of lawyers "whose trade it is to question everything, yield nothing and talk by the hour." But he listened patiently. In time he presented his own plan, his voice deep and gentle, his proposals brief and blunt. He illustrated his remarks with a map which showed the open territory as far west as the Mississippi divided into fourteen new states.

Jefferson hardly regained his seat before the attack began. The congressmen balked at his boundary lines. They scoffed at his suggested state names. They grew shrill and scornful over his plan that slavery be forbidden in the new states.

But in the end Jefferson succeeded in winning some important principles. Congress agreed that new states would be formed as quickly as their areas achieved 60,000 free inhabitants. It was decided also that the newcomers into the union "shall be admitted on an equal footing with the original states."

Eight years later Vermont would enter the Union as the first of the new states. Similarly, Kentucky, Maine, and later West Virginia were formed by splitting away parts of some of the original thirteen states. Tennessee and Ohio had enough population to be admitted directly to statehood. However, the additional undivided lands within the boundaries of the original nation were organized as territories before they became states.

To Jefferson's way of thinking the seaboard was a jumbled patchwork of boundaries that jarred his sense of order. The borders were as twisted and confused as those of the Old World.

BRITISH TERRITORY

THE UNITED STATES
After the Treaty of 1783

VIRGINIA

MASSACHUSETTS CLAIM

CONNECTICUT CLAIM

CLAIM

VT. Claimed
by N.Y. and N.H.

N. H.

MASSACHUSETTS

Mass.
Claim

NEW YORK

CONN.

R. I.

PENNSYLVANIA

N. J.

MD.

DEL.

VIRGINIA

KENTUCKY

NORTH CAROLINA

NORTH CAROLINA
CLAIM

TERRITORY

S. Carolina Claim

SOUTH
CAROLINA

GEORGIA CLAIM

GEORGIA

Claimed by Spain,
United States, and Georgia

ATLANTIC OCEAN

GULF OF MEXICO

Political units and public land holdings in the thirteen original states were of random shapes and sizes. Their edges were as zig-zag as the courses of wandering creeks and just as difficult to measure or survey.

Jefferson realized that these lines could no longer be changed without bloodshed. But he now set to work on an orderly system of land division and measurement for America's new states. His plan, which Congress eventually approved, called for the ar-rangement of the public domain into squares. The basic unit would be a six-mile square known as a township.

The new law sent hundreds of surveying teams out into the wild realms of the buffalo, the eagle, and the Indian. They dragged their measuring chains over the silent hills and the shim-mering prairies. Six miles the township lines ran, then turned at right angles for another six.

The government surveyor's chief instrument was the compass. All the lines ran north-south and east-west, like the parallels and the meridians. The first teams began in 1705 at the point where the Ohio River meets Pennsylvania's western border. Measur-ing their way across Ohio, their job was to "lay out straight lines across crooked land."

Soon the crews were hacking and clearing in the piney forests of Alabama. The work moved fast across the open prairies of Illinois, where a man might see all four corners of the township from a single point. The wage was a dollar a day, paid alike to mountaineers in Kentucky, to axmen in Wisconsin, and to chainmen wading through the Mississippi swamps.

They reported to a head surveyor in charge of a range, which was a row of townships. The information was then transferred to maps. Gradually vast areas of government-owned wilderness were neatly plotted out in standard squares and tied into the world-wide grid system of measuring latitude and longitude.

In their sketches and notes the surveyors made a detailed record of the countryside. They set down the exact locations of wellsprings and Indian trails. Reported were the evidences of underground minerals and the depths of lakes. Their jottings even told about the feel of the soil and listed the species of trees.

Each square township was split into thirty-six equal parts— one-mile squares. These came to be known as sections, each comprising 640 acres. Under the law one section near the center of each township was turned over to the public schools.

The government surveys resulted in a chain of effects. Large areas of wilderness were mapped with relative accuracy. Opening land offices as soon as an area was mapped, the government was able to sell its land to settlers. The migration began in earnest. Pioneer trails became crowded with eager families holding maps and deeds, searching for the blazed tree or the stone column which would mark the corner of their new homesteads.

However, the deed promised neither success nor safety. A thousand perils faced the homesteader. Countless frontier homes were destroyed by Indian firebrands.

The government rule was that only those lands would be sold which had been secured from the Indians by treaty or purchase. However, this principle often had little practical meaning. The Indians believed deeply that the only things that could be sold were items one could carry away. Land could be taken only by conquest.

Nevertheless, government agents "bought" large pieces of the wilderness at the lowest price they could get. Often this meant only that a tribe was too weak or too hungry to resist an offer. In one of the largest land deals in the Middle West the Indians "sold" property at six acres for a penny.

Working in Indian country, surveyors were often escorted by government troops. But many were later found scalped in the underbrush. Where the surveying parties went, the Indians noted, there was a swift change in the hunting lands and fishing streams their tribes had used for generations. The Indians would return one spring to find the trees felled and the land plowed—and themselves on the outside of a fence.

The red men continued to fight their losing battle for the land. In the end they would become a handful of outcasts in a country that was once theirs. Ironically, they are commemorated by the Indian names of many states and the Indian place names dotting the map of America.

As the century ended, Thomas Jefferson could look with satisfaction at the orderly growth and settlement of the country. He had been elected President of the United States. On the map he could see the new state and county borders fixed forever in the pattern of the nation.

All the way to the Mississippi River the surveyed square townships had their effect on the map of the land. Most of the counties were laid out in squares and rectangles. The roads and streets, the fences and hedgerows tended to follow the main compass points. Many new towns were planned in the crisscross patterns of straight lines and right angles.

Jefferson was now thinking about the land beyond the Mississippi and how it might become part of the United States. The President already had his envoys abroad quietly discussing the Louisiana territory. With patience, Jefferson believed, the tangled politics of Europe might help bring this prize into American hands.

The Louisiana territory, largely unmapped and unexplored, excited Jefferson with its possibilities and its mysteries. This learned man had only the vaguest notion of what wonders might be found in that wilderness. He had heard of mountains in the West, but it was still his opinion that none was higher than the

Peaks of Otter which he had personally surveyed in the Blue Ridge range. He also believed that curved-tusked mammoths, actually long extinct, were at that moment thrashing and trumpeting about in the western forests.

Into Jefferson's thoughts frequently came the figure of a young man who had been his neighbor and friend back in Virginia. He was a bright, blue-eyed, sturdy youth named Meriwether Lewis. The two had often gone on long rambling walks, talking about America's future and the great western frontier. Both Jefferson and young Lewis shared an interest in tinkering with scientific things. They had even rigged up a system of mirrors by which Jefferson could summon the boy across the Virginia hills and farmsteads. At that time young Lewis had told the man who was later to be President: "Call me whenever you need me."

It was not long after he was inaugurated that Jefferson summoned Meriwether Lewis to the White House. The young man was now an Army officer with a distinguished record of frontier service.

There were some in the capital who were puzzled as to why Lewis should be chosen as Jefferson's private secretary. The young man had many good qualities, but he was not especially trained in White House tasks. Lewis was clearly more at home in the field than he was in the office, and his spelling was rather poor.

However, Jefferson and his secretary had a secret between them which they often discussed in the President's map-lined study. Before long, word leaked out that the secretary was to have duties no secretary had ever had before.

5.
Two Virginians

In the year 1803 this newborn nation doubled its size through the Louisiana Purchase. At three cents an acre the United States bought a vast territory stretching from the Mississippi River to the Continental Divide.

France, needing the money more than the land, hardly knew what she had sold. The United States had only the vaguest knowledge of what she had bought. No white man had ever explored the country to the Continental Divide, much less the overland route to the Pacific beyond. To President Jefferson fell the task of naming the man who would find and map his way across the trackless wilderness.

It was a starlit May night in 1803. With the aid of a few instruments, Meriwether Lewis was finding his whereabouts—40° north latitude, 75° west longitude. Far from being lost, he knew exactly where he was—in the city of Philadelphia. He also knew his future course. In his pocket was a letter from the President of the United States, carrying secret but clear instructions:

"The object of your mission is to explore the Missouri River, cross the highlands, follow the best water-communication to the Pacific Ocean."

Lewis was a tall, strongly built man with corn-colored hair. He was serious and even moody at times. Men sensed in him a deep well of patience and strength. "Captain Lewis is not regularly educated," Jefferson was telling his friends, "but he is brave, prudent, habituated to the woods and familiar with Indian manners and character."

Weeks and months of careful preparation were ahead for the young man, still in his twenties. The excitement of the journey had already begun.

Lewis's four-week visit to Philadelphia was aimed at quickly gathering scientific information which might be useful on the expedition. This was America's center of science, full of President Thomas Jefferson's learned friends. At the President's request, a group of them had taken the young man in hand. Botany, zoology, astronomy, mineralogy—the tutors dealt out their concise information just as fast as their student could absorb it. Indian history, health measures, and quick remedies for frostbite and snake bite, all these were covered in rapid fashion.

The young man was shown how the sun and the stars appear to move across the sky and how man locates himself by observing these movements. "When necessary," he was told, "man brings the heavens down to the earth." Lewis was taught to think in terms of degrees, to split each degree into sixty minutes and each minute into sixty seconds. His days were spent under Philadelphia's blossoming nut trees, mastering the use of the surveyor's instruments. Each clear night he was out studying the stars. Lewis's task would be not merely to explore a pathway to the sea but to chart and map and survey it as well.

The dawning nineteenth century boasted some improved instruments for the measurement of latitude and longitude. And Lewis would have the best available. He acquired a sextant, a triangular brass instrument for sighting objects in the sky and measuring their height above the horizon.

The long struggle to solve the problem of longitude had ended with the invention of the chronometer, an accurate, portable timepiece. The meridians that are 15 degrees apart in longitude are one hour apart in time. Longitude could thus be found by carrying an accurate timepiece showing what time it was at the prime meridian in Greenwich, England. The traveler simply calculated the difference between local sun time and Greenwich time, converting each hour into fifteen degrees of longitude. Into Meriwether Lewis's hands went a fine gold chronometer, together with information on how to use it.

Throughout Jefferson's letter of instructions ran an emphasis on the charting of the long route. "Take careful observations of latitude and longitude," wrote the President, "especially at the mouths of rivers, rapids, islands." And Lewis prepared himself for his mapmaking duties by every means that he could.

The man Lewis chose to accompany him on the trip and to share command was William Clark, a Virginia friend. Both had served together as young militia officers. Back in Washington in June, Lewis had written to his friend explaining the project in which he was already so deeply involved.

"If there is anything . . . which could induce you to participate with me in in its fatigues, its dangers and its honors, believe me there is no man on earth with whom I should feel equal pleasure in sharing them as yourself," Lewis wrote.

The slow mail brought Clark's reply a full month later: "No man lives with whom I would prefer to undertake such a trip."

While growing up. William Clark had suffered the fate of every boy who has a famous older brother. Before William was full-grown, his brother, George Rogers Clark, was already a popular hero as a frontier fighting man. The elder Clark helped secure the new nation's narrow borders; the younger brother would help extend them from sea to sea.

Born in Virginia, William Clark had been raised on the Kentucky frontier. His resourcefulness in the wilderness would carry the transcontinental expedition through some of its greatest trials. Clark had a good knowledge of geography and engineering. Both friends, Lewis and Clark, were experienced in the outdoor life, in frontier travel, in leadership. Together they made a superb team.

There was no longer any need now for secrecy about the trip. The unmapped Louisiana Territory to be crossed by the Lewis and Clark Expedition was now firmly in American hands.

The two young Virginians made their preparations. They gathered together a crew of 29 woodsmen, rivermen, mountain men. Included were a one-eyed steersman, a black slave, an interpreter skilled in Indian languages, and a blacksmith with a talent for repairing almost anything.

The company assembled in the damp, cold closing weeks of the year to winter together on the Mississippi River bank near St. Louis. There the party became familiar with their large keelboat and the two double-pointed pirogues—and with each other.

Under Clark's leadership that winter this varied assortment of men went through rigorous training and became a unit. They called themselves the Corps of Discovery. For the next two years they would be bound together in an adventure wilder than they had ever dreamed. Again and again they would save each other from danger and death. And much as they might grumble at the hardships and cry out against gnawing hunger, there would never be a major quarrel among them.

In the spring they watched the debris-filled water of the Missouri come rolling out of the West—just as Louis Jolliet and his party had seen it more than a century before. On May 14, 1804, the Lewis and Clark Expedition was under way in a driving rain, the great square sail of the keelboat flapping in the prairie wind.

From the outset disaster followed in the wake of the party, but it never seemed to catch up with them. Squalls and sand bars threatened the little fleet. The unpredictable river changed course to every point of the compass. The Missouri overran its banks and bolted its own channel. At times the water turned to seething fury, commanding the full strength of every man at oar and pole. The current forced the huge keelboat under an overhanging tree, smashing the mast to bits.

Often the crew was on the riverbank, dragging the craft upstream with towlines. The expedition moved slowly west and north, averaging a dozen miles a day.

On a sad morning in August the voyagers buried one of their men, Charles Floyd, who had died after a lingering illness. They dug his grave on a high bluff and marked it with a cedar post. Lewis put the nearby stream on his map and named it the Floyd River. It can be found on today's maps flowing past the place where Sioux City, Iowa, now stands.

Nearby were cliffs where Indians quarried the red stone for their peace pipes. Lewis soberly noted that to the Indians the whole region was sacred, and even warring tribes might meet there without any show of violence.

With Clark in charge of the boats, Lewis spent most of his

time along the riverbanks, his black dog Scammon at his side. There were endless new wonders of nature to be explored. The voyagers delighted in the song of a whippoorwill which perched on the bow of the keelboat, and they tamed a beaver as a shipboard pet. Astounded, they checked the carrying capacity of a pelican's pouch, pouring five gallons of water into it.

Lewis observed and smelled and listened and tasted. When he sampled some blue powder, he was sickened at the taste of cobalt. He noted carefully Indian customs, legends, languages. Many a day ended with Lewis sitting on the riverbank, writing in his diary and slapping at mosquitoes, putting the pests into the record as well.

To Lewis the darting of the shy pronghorns seemed "more like the flight of birds than the movements of earthly beings." He was amazed to watch the herd pass swiftly before him and suddenly reappear on a distant ridge.

The careful mapmaker returned nightly to his study of the stars. But Lewis responded also to the spangled beauty of the heavens over the sunflower prairie. He recorded that one dawn "the morning star appeared much larger than usual."

Clark participated fully in the mapmaking work. He seemed to have some special feeling for the landscape, a talent for figuring out in advance what was ahead but still unseen. In his map sketches he had a knack for catching the wrinkle of the hills and the paths of the watercourses.

On the day of the autumn equinox, thick fog hampered Lewis's celestial observations. But he estimated the party to be at 44 degrees, 11 minutes, 33.3 seconds ($44°$ $11'$ $33\frac{3}{10}''$) north latitude. It was time to think of setting up winter camp. But the expedition was now in Sioux territory, among Indians who were warlike and dangerous. In a series of close encounters, the party managed to evade bloodshed. The two leaders were steadfast in carrying out their exploring mission, and they firmly avoided trouble whenever possible.

However, to many tribes the full purpose of the white man was now becoming clear. And they were making their stand against the invasion from the East. The Indian envied the newcomer's guns, his horses and tools, but he had little respect for the white man's fairness or honesty or wisdom.

Wherever they went Lewis and Clark noted the effects of the white man's smallpox, gunpowder, and whisky. Some of the tribes shunned the white man's ways and his gifts. The Arikaras rejected offers of whisky. Lewis noted in his journal: "They say we are no friends or we would not give them what makes them fools."

On a November night the company halted near the villages of the light-skinned and friendly Mandan Indians. The explorers built a winter encampment near what is now the North Dakota capital at Bismarck.

Lewis calculated from his map that the Corps of Discovery had gone 1600 miles from the mouth of the Missouri River. That map, which Lewis had prepared for the journey, was a strange one. It was a map of America's West. Its eastern edge was fringed by the Mississippi, with a few sketchy features to the north and south and the outline of the Pacific shore on the west, as it was known from coastal maps drawn by a number of British and Spanish explorers.

Across the wide sweep of the map was nothing but white space —blank, except for a thin and lengthening line showing the progress of the expedition. Each day Lewis added to that line.

A full year passed without any word back home from Lewis and Clark. And then one spring day a number of odd-shaped bundles arrived at President Jefferson's half-built White House.

BRITIS

OREGON

Ft. Clatsop

Lewis

Columbia River

Clark

COUNTRY

Yellowstone River

Ft. Mandan

Three Forks

Snake River

Obtain horses

ROCKY MOUNTAINS

LOUISIAN

SPANISH TERRITORY

Rio Grande

RRITORY

LEWIS AND CLARK
EXPEDITION
1804-1806

Missouri River

St. Louis

Kansas City
(Independence)

Ohio River

UNITED STATES

Mississippi River

RCHASE—1803

SPANISH TERRITORY

On their keelboat the explorers had shipped back animal skins and bones, Indian relics, and living specimens of western wild life. Meriwether Lewis recorded that he had sent "a box of plants, another of insects, and three cases containing a burrowing squirrel, a prairie hen, and four magpies all alive."

Included among the packages was a rolled-up map showing the party's route and the terrain well into the Louisiana Territory. The President was delighted. The map revealed clearly that Lewis and Clark were making good progress.

In place of the bulky keelboat the river voyagers had built six wooden canoes out of green cottonwood. Lewis noted in his journal:

> This little fleet although not quite as respectable as those of Columbus or Captain Cook, was still viewed by us with as much pleasure as those deservedly famed adventurers ever beheld theirs. . . . We were now about to penetrate a country at least two thousand miles in width, on which the foot of civilized man had never trodden.

That spring the expedition moved across the northern countryside in bitter cold weather. Some mornings the men shook the snow from their blankets and found thick ice in the water kettle.

The journey now lay steeply upstream in the narrowing river. The men were spending a good part of their time in icy water up to their armpits, hauling the boats against the current, guiding them through shoals and rapids and around willow islands.

The Corps of Discovery had added three new members: a French-Canadian guide, Toussaint Charbonneau, his seventeen-year-old Indian wife, Sacajawea, and their newborn baby, which Lewis had helped bring into the world.

The company learned the full tragic story of the young Indian woman. While still a child, she had been taken by an enemy tribe. Wandering far from her mountain home, she had become the wife of the trapper Charbonneau. Her hope was to be able to visit her own people.

For days Lewis and Clark seriously discussed the hazards of taking a woman and an infant on the perilous journey. They realized that Sacajawea might prove to be very helpful. Perhaps she could assist in procuring from her tribe the horses they would need in order to cross the mountains. Moreover, Sacajawea's presence would show Indians that this expedition was not warlike. Lewis said finally: "A woman with a party of men is a token of peace."

The two Virginians, aware that the most difficult part of the journey was ahead, realized that they depended heavily on the good will of the Indians. Throughout the trip, the two of them even remained smooth-shaven so as not to seem too strange to Indian eyes. However, the sun-bronzed pair made an unusual appearance, the blond Lewis and Clark with his flaming red hair.

The Indians were astonished at the skin coloring of York, the Negro member of the party. They marveled at the sound of the violin played by one of the boatmen. And they peered with interest at Clark's sketched maps, so different from the maps which the Indians occasionally painted on hides or formed from sand or campfire ashes.

The entire party was now dressed in deerskin moccasins, leggings, and shirts, fringed so that the rainwater dripped off the garments instead of soaking into the seams.

Around them spread a vast country, lonely and dark except for a few scattered Indian fires. The wayfarers were now close to the heart of America's deepest mystery. For three centuries men had puzzled over the crossing of the continent, pondering the possibility of some easy passage by land or water.

The mystery was locked deep in uncertainty as to the breadth of the continent and the shape of its terrain. The western mountains had appeared on maps drawn by men who had never seen them. Indians called them "the shining mountains." And on maps they were often labeled "the stony mountains." There were many who believed these mountains formed a single high but narrow ridge.

Others believed that the east- and west-running rivers flowed from a single source somewhere in the highlands. President Jefferson saw the Missouri River as reaching up to the dividing crest of the continent, with the source of the Columbia River just beyond, running down to the sea.

On the Lewis and Clark map, the truth was slowly being revealed. The voyagers had reached the western edge of America's central plain. Ahead was the mountain barrier, higher and wider than any of them had dreamed. The peaks loomed up, rank upon rising rank, topped by snowy pinnacles that pierced the cloud banks. These mountains were neither a slender ridge nor a single range, but an awesome network of towering chains and systems.

The expedition was in virgin country, uncharted and unnamed. And on their map newly chosen names of tributary streams, islands, and peaks began to appear. With very little ceremony the Corps of Discovery left place names that would remain forever.

At one point, they encountered a river whose water had "a peculiar whiteness such as might be produced by a tablespoon full of milk in a dish of tea." Lewis added, "We called it the Milk River." They named features on the landscape for members of the party. And one evening Clark, now homesick and thinking about his Virginia sweetheart, a certain Miss Hancock, made a special request. Lewis complied by naming a large south-flowing stream the Judith River.

On June 3, the Missouri presented the troop with a crucial

problem. The river divided itself into two branches of almost equal size. Which was the real Missouri? From this time on, men would have maps to answer this question. But for Lewis and Clark the riddle was a real one. They knew only that the correct choice would lead them to what the Indians described as a series of high waterfalls.

In the following days the entire company pondered the problem. They split into two squads, explored the forks, heatedly argued over the choice. Lewis, with a group inspecting the right branch, failed to convince his squad that this was *not* the Missouri. That night the commander lay restless in his wet blankets, weighing the decision. A mistake could lose valuable weeks for the expedition, perhaps stranding them in the mountains in the dead of winter.

Lewis chose the left fork, convinced that the right one had its source in Canada. The explorers plunged on now through torrential rains and a series of dangerous incidents. Sacajawea had become seriously ill and had to be carried over the rugged, trailless terrain. The quiet Indian girl, warmhearted and loyal, had endeared herself to the entire company. She ran a high fever, and the men feared she would die.

Lewis led the overland caravan through slippery, rocky country, each man walking in the footprints of the man ahead. Several times the men lost their footing and came within inches of death. Here the party fought rattlesnakes. A grizzly bear reared up eight feet tall and charged furiously even after a half-dozen bullets had found their marks.

It was a disheartened company now, sick of hardship and bone-weary. But late one evening they heard what sounded like thunder. Lewis described it in his diary: "My ears were saluted with the agreeable sound of a fall of water and, advancing a little further, I saw the spray arise above the plain like a column of smoke. . . . I hurried down the hill to gaze at this sublimely grand spectacle . . . the grandest sight I ever beheld."

The joy at seeing the impressive beauty of the Great Falls was heightened by the knowledge that they had chosen the correct fork as the Missouri River. However, the falls soon confronted the party with the biggest problem of travel they had yet encountered. But Clark's specialty was on-the-spot inventiveness. Brimming with good humor, this ingenious leader responded with a plan.

Fatigue was put aside and the men set to work preparing for the long overland haul around the falls. From the trunk of a huge cottonwood tree they sliced sections to be used as wheels for a number of wagons. The masts of the pirogues served as axles. With their equipment piled aboard these makeshift vehicles, the men put their shoulders behind them, while axmen ahead cleared the way through the thick brush. A month and a half was spent in this backbreaking portage.

The narrowing Missouri soon divided into three branches. Lewis named them for three leading members of the United States government, calling the largest of them the Jefferson River. As they climbed to the Missouri River's source, one member of the crew marveled at how the great river had dwindled to a trickle among the stones. Back home he would be able to boast that he had "walked across the mighty Missouri without getting his feet wet!"

The Lewis and Clark Expedition was now approaching one of the most important dividing lines on the map of America.

Running from north to south is the Continental Divide, the great watershed of the West. It is the spine of North America, separating drainage running toward the east and toward the west. The future course of a raindrop, a snowflake, a hailstone may be determined by which side of the Continental Divide it strikes after its descent to earth. If it falls on the east side, the bit of moisture may be carried by streams all the way to the Mississippi and into the Gulf of Mexico. Moisture dropping to earth on the west side of the Continental Divide may find its way to the Pacific.

Late in that summer of 1805 the explorers struggled up to the crestline of the continent. One evening they watched a dried leaf drift down to a stream at their feet. They noted that the current carried the leaf westward toward the setting sun. They had now crossed the Continental Divide, and from here on their river journey would be downstream.

The two leaders realized that the narrowness of the mountain barrier was a myth and the easy passage to the sea was a dream. Lewis had seen enough of the Rockies to know that much difficult travel lay ahead of the expedition. His main concern was with horses and with finding Indians from whom they might be secured.

Weeks had passed since they had seen any human being other than members of their own party. And now they were scouring the mountains for any cold ashes, any trail, any smell of smoke that might lead them to an Indian camp.

One morning Lewis and two men moved out ahead in search of Indians. They traveled far and fast. After several days they spotted a single Indian on horseback, only to have him shy away when they approached. Days later the trio encountered a few Indian women who ran in fear of the strangers. Only by patient and repeated efforts at friendly approaches did the white men succeed in making contact with the Indian camp and in reaching their chief.

These were Shoshonis, a poor and hungry tribe repeatedly defeated in many battles and raids. Lewis could not convince them of the peacefulness of his mission, of his need to buy horses, or that the rest of his party were some days distant and would soon arrive. The three white men were received with great suspicion. The young chief was afraid they were allied with enemies.

To Lewis's great relief, Clark finally appeared with the rest of the party. The white men now observed strange behavior on the part of Sacajawea. She obviously recognized several of the Shoshoni women as childhood friends. Some of these women had been taken prisoner along with her and were later separated. Now the Indian women held each other closely and expressed their joy in warmhearted fashion. Sacajawea, long lost and feared dead, was welcomed back into the bosom of the tribe.

Lewis brought Sacajawea before the chief to act as interpreter. But to Lewis's astonishment the two embraced, weeping for joy. At length the bewildered Lewis was able to get an explanation from Sacajawea. The two were sister and brother!

When Lewis and Clark left the camp of the Shoshonis, they had twenty-nine horses and a mule secured through trading. They also had the services of a veteran Indian guide called Toby, who would lead them through the mountains. Sacajawea and her husband made the decision to continue on the journey with them.

The September days were growing shorter and the expedition was now in a race against the onset of winter. They threaded their way across hazardous crags. Several times heavily loaded horses tumbled to their deaths down the steep slopes of what Lewis described as "these terrible Bitterroot Mountains."

They had been able until now to depend on the rifles of the sharpshooters in the party for fresh game. But now the hunters came back each day empty-handed. The explorers, already hampered by fatigue, were now weakened by hunger. A bitterly determined Lewis wrote in his journal: "If the Indians can survive here, so can we."

The foul weather added to the difficulties. It seemed to the men that they could never get fully warm or dry. This was one of the many times when the Corps of Discovery moved ahead on Clark's cheerful courage and Lewis's high sense of duty.

On a snowy morning Lewis awoke early and shivered as the cold mist descended the mountain. With his sextant he took a reading on the sun. The Virginian blew on his hands and coaxed the frozen ink to flow from his pen.

The map spread out before him showed that the expedition's line of travel extended far to the west. Only a short gap of blank space remained between their position and the Pacific shore. But now the barrier of the coast mountains rose a little higher every day. And Lewis could only guess at how many weeks of travel lay ahead. Would the year's end find them at the ocean —or snowbound and starving in these same mountains?

Lewis raked up the dying fire. His companions lay sprawled about on the ground in their blankets. No night's rest, no week of sleep would end the dull ache of weariness in their bones. But looking back along the mapped route Lewis knew that this small body of men had already added greatly to the meager fund of his government's knowledge about this country.

In the rosy light of the sunrise Lewis could trace the line where their travels had changed groping guesswork into charted reality. Hundreds and then thousands would follow in their track. But the young commander could hardly imagine the extent of the western migration which was to come. Gold rush and land rush, railroad and airplane would eventually bring the people streaming across these mountains. Nor could he envision the towering cities which would rise along their trail—Kansas City, Omaha, Bismarck, Great Falls, Portland.

Occasionally the caravan caught sight of west-running streams. But these were wild mountain torrents which would certainly dash a boat to bits. It was not until mid-October that the party was able to risk river travel again. In newly built dugout canoes they sped down the perilous cascades. They were fighting the calendar on their way to the sea—but at times they wondered if they would arrive whole.

In the lead canoe was Pierre Cruzatte, the one-eyed veteran boatman who skillfully piloted the fleet through dangerous rapids. A series of wandering streams led them at last into the plunging waters of the Columbia River. Here they passed between high canyon walls and caught vistas of sheer peaks, the highest they had seen.

Their river route would someday be the dividing line between the states of Washington and Oregon. However, this was still disputed territory claimed by Spain and Britain, and partly by Russia. On the strength of Lewis and Clark's exploration the United States would make the strongest claim of all.

Along these banks were villages of Indian families, salmon fishermen who were astonished at the strange fleet that sped downstream, riding the wild, bucking current.

Lewis knew now that nothing could stop them until they reached the crashing white fury of the ocean surf. But each night when the expedition made camp he tried to estimate the distance to the Pacific. He was like Columbus, who, crossing the Atlantic more than three hundred years earlier, had searched westward for any sign of land. Lewis now sought eagerly for any trace of the ocean. His heart leaped at the sight of an Indian wearing a British seaman's jacket. Flights of sea birds passed overhead. He noted Indian ornaments made of sea shells.

On November 7 Lewis led the main party overland to an Indian village in the mountains. Beyond a crest a vista appeared which set the entire company cheering in jubilation. Through the mist below they could see a rock wall holding off the thundering, majestic Pacific Ocean.

In that wild moment the men forgot their empty stomachs and their weariness. They shouted and danced in delight. The mountains around them echoed to the sound of Cruzatte playing a joyous jig on his violin.

The terse entry Lewis set down in his journal that momentous day read: "Great joy in camp. Ocean in view!" The line on the map was complete.

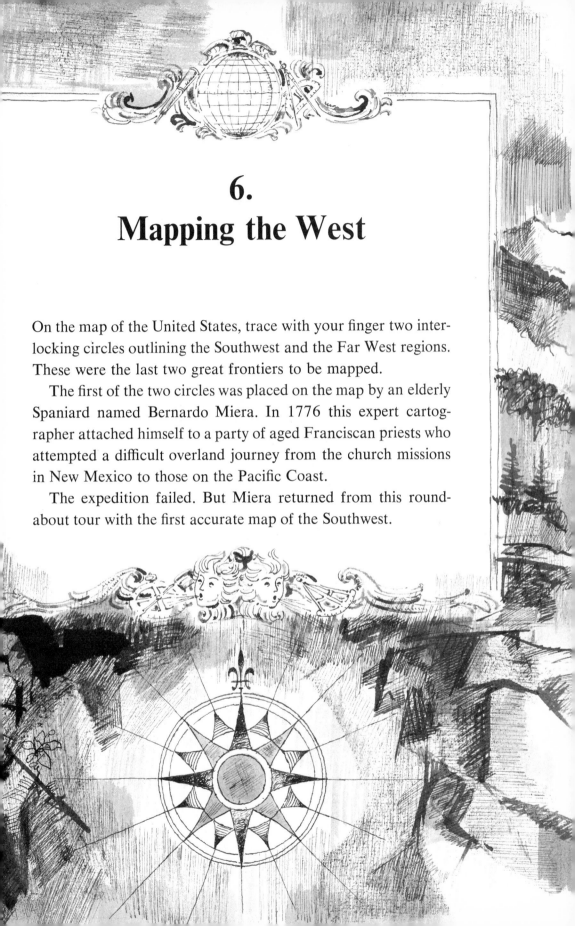

6.
Mapping the West

On the map of the United States, trace with your finger two interlocking circles outlining the Southwest and the Far West regions. These were the last two great frontiers to be mapped.

The first of the two circles was placed on the map by an elderly Spaniard named Bernardo Miera. In 1776 this expert cartographer attached himself to a party of aged Franciscan priests who attempted a difficult overland journey from the church missions in New Mexico to those on the Pacific Coast.

The expedition failed. But Miera returned from this roundabout tour with the first accurate map of the Southwest.

The second circle was mapped out by a man who was totally unlike the exploring friars. He was a young, roughhewn American fur trapper named Jedediah Strong Smith.

Lewis and Clark were the idols of many American boys. Among them was a spindly Yankee youngster who was stalking squirrels along the Susquehanna the year that the two Virginians returned from the Pacific.

He was christened Jedediah, but he seldom answered to that name. People called him 'Diah and Jed and, later, Old Smith—although he didn't live to be very old.

His family doctor gave young Jed Smith a book which recounted the adventures of Lewis and Clark. The explorers and their historic overland crossing to the sea never again left the boy's thoughts. Jed Smith spent his growing years trying to figure out some way that he could follow the track of Lewis and Clark. Poor, uneducated, and without influential friends, he was not likely to get backing for an exploring trip into the wilderness. As he studied copies of the maps which the pathfinders had made, the West seemed not only distant but unreachable.

But in time he was to draw his own great map. It would link the maps of Lewis and Clark with those of Miera and Coronado in the Southwest. Jed Smith would learn more about the West than any other man of his day. He was to become one of the most remarkable of the American mapmakers, working without instruments, knowing nothing about observing the heavens. Instead he determined his position by means of old maps and by his own sixth sense of geography. He roughly estimated distances and guessed at directions. And yet his maps came closer to the truth about the West than those of any man before him.

A newspaper advertisement that appeared in 1821 caught young Jed Smith's eye and set his heart hammering against his ribs. It read:

> To enterprising young men. The subscriber wishes to engage one hundred men to ascend the river Missouri to its source, there to be employed for one, two or three years . . .

Jed Smith was soon in St. Louis, with the newspaper in his hand and his hopes mountain-high. He was twenty-three, with wiry muscles, a keen intelligence, and a sharp eye to his rifle sights. Those qualities were enough for him to be hired as a member of a band of grizzled fur trappers who made their way up the Missouri that year.

At campfires along the broad river Jed learned the hazards of the trapper's life. Few lived to old age. But there was money to be made. With luck a man might pile up a fortune in a few seasons, then return home to enjoy his wealth.

Jed Smith was introduced to the Western wilderness in the embrace of a grizzly bear that tried to bite his head off. After a fierce struggle, in which the huge beast was finally killed, Smith lay horribly chewed and clawed. While his companions stood by appalled, Jed coolly told them how to dress his wounds and sew back a dangling ear with needle and thread.

He wore the ugly scars for the rest of his life. But scars were the trademarks of the fur trapper. Jed was now a member of a reckless breed, deadly with the long rifle, skilled with traps, and as wary as any creature in the wilderness. The trapper lived off the land, slept with sleet in his face, and seasoned his food with gunpowder.

The animal furs he gathered were in high demand. Buckskins sold for a dollar, so that the dollar became known as a "buck." Buffalo robes were used as comforters at many an American hearth. But the most popular furs of the day were the beaver skins, which were shaped into men's hats. Beavers were plentiful along the riverbanks all across the northern regions of the continent. And rivalry was keen. The fur-trading companies fought bitterly over favored hunting grounds.

The trappers found themselves in the front line of the continuing warfare between white men and Indians. The clashes were fierce and frequent. Often they killed each other on sight, robbing and cheating when they could. No man is as desperate as the one who feels he has been tricked out of his livelihood. And this was the mood of the Western Indian as he watched his hunting and fishing grounds being taken away before his eyes. On the other hand, the trapper saw the Indian as one more deadly peril of life in the wilderness.

Three years of working trap lines in the central Rockies brought out in Jedediah Smith qualities of leadership and extraordinary courage. A pair of inner forces seemed to drive him on tirelessly. One was his burning ambition to succeed as a fur trader. The other was his deep curiosity about the American wilderness.

He was struck with scenic grandeur few white men had ever seen. He was now making notes and map sketches and writing long descriptive letters about the magnificent natural setting in which he found himself. He wrote about the great buffalo herds. It seemed to Smith that "all the buffalo in the world were running in those plains . . . making the ground tremble with the moving weight of animal life."

Smith made a careful exploration of the South Pass through the Rockies. This important cleft in the mountain barrier which he mapped was to become the main route of the pioneers moving along the Oregon Trail.

He was troubled when the published maps of his day did not correspond to the Western landscape as he saw it. Many of these maps were a jumble of errors, made by men who did not have the trappers' firsthand knowledge of the terrain. However, the trappers usually had neither the desire nor the skill to make maps. Jed Smith was the exception. From his companions who scoured the mountains and plains he gathered a rich store of knowledge to add to his own.

The great piles of furs which he and his party shipped back to St. Louis each year made Jedediah Smith an important figure in the fur trade. But in the summer nights of the year 1826, a strange and nameless need still stirred inside him. Secrets hidden beyond the silent plains and the shrouded peaks would not let him rest.

There were mysteries in this western land. The names of them echoed in his brain. Buenaventura—this fabled river that men had been putting on maps for a century—but who had ever seen it? And California, a land of legend held by Spaniards—but they seldom ventured to explore it. Smith had heard about a strange inland lake, saltier than the sea. Somewhere there was a desert valley from which no man emerged alive. Word had come to him also about a canyon a mile deep, cut into the bedrock by a river that ran in a white rage.

He could not fully understand what lured him toward the unknown. A year later he would try to explain it in a letter to his boyhood hero, William Clark, now a high U. S. government official in St. Louis. "I started out," the letter ran, "for the purpose of exploring the country southwest which was entirely unknown to me, and of which I could collect no satisfactory information from the Indians."

By August, Old Smith was on his way south. With a rifle across his saddle, he was a tall figure in buckskin. His sharp features, burned coppery by sun and wind, gave him an Indian appearance. He was smooth-shaven, with long dark hair. The only head covering he wore was a blue bandanna.

Strung out behind him was a caravan of horses and pack mules. His party numbered a dozen men, all devoted to Old Smith, but baffled by his plans. Their leader was a man of simple speech and little humor. Unlike most trappers, Smith didn't drink or gamble. The only thing he ever smoked was an Indian peace pipe. But Jed Smith never preached his religion to anyone. He realized that he was in a country without law, among men who made up their own rules of living. He simply shrugged when mountain men sometimes went as wild as the animals they hunted. In turn, the trappers tolerated Old Smith and his strait-laced ways. They respected him as a tough, resourceful leader.

Out from the shadow of towering peaks, the company rode into the hot buffalo plains. The Indians they met here were un-familiar with white men and consequently cautious. They accepted the trappers' gifts of knives, combs, needles, mirrors, awls, and buttons. In turn they gave information about the terrain ahead.

Day by day the landscape became more dry and desolate. Jed Smith was to find this "a country of starvation." To the trappers the land looked hopeless for games and hostile to man.

They traveled for weeks through a wasteland. Men would later say of this region that its only useful purpose was to hold the world together. But there were wonders which Smith noted on his map sketches—salt caves, twisting canyons, high tablelands, natural bridges. Red earth and sagebrush ran out on every side as far as the eroded hills. The leader still pushed to the south-west into the torrid setting sun. Jed Smith's men were weary and bewildered.

The end of summer found the party with few skins, carrying their rusty traps. The men whispered questions among them-selves. Has Old Smith gone mad? Why is he driving deeper and deeper into a country without game? What is a fur-trapping party doing in a region where the largest fur-bearing animal is the desert rat?

But Smith moved on in the sure knowledge that he was explor-ing new country. Estimating from his own maps, the expedition had reached California. Before him now was a vast, barren desert which would someday be named for the Indians who lived on its edge, the Mojaves.

Through the agony of the next days the party lived under the eye of a fiery sun that took a heavy toll among the horses. The men too sensed the presence of death. Against the horizon stood the weirdly twisted silhouettes of the Joshua trees. Sundown brought only the mournful howl of the high wind.

Covered with dust and sand, the company trudged at last into a Spanish mission near what is now the city of Los Angeles. The padres and the Spanish officials could not believe their eyes when they saw Jed Smith's party emerge like ghosts from a desert they had thought to be impassable.

Smith found no hospitality in California. Mexico, California, and the entire Southwest were seething with revolution. Spain was losing its grip on the continent as the Mexicans rebelled and took over large regions that had been Spanish-held. Under these circumstances there was no welcome for these first Americans to reach California by an overland route. Instead Jed Smith was treated like a common outlaw or an American spy.

When he and his men were finally released after weeks of captivity, Smith was rudely ordered out of the country. But the stubborn trapper would not leave until he had traveled the length of California, exploring its broad central valley. It was then that he turned east—only to run into the greatest barrier of all, the Sierra Nevada.

Old Smith was in the lead as the party began to climb the lofty summits of this saw-tooth range. But horses and men balked at the icy trailless heights. Heavy snow blocked their way and blinding blizzards made travel impossible. The horses plunged chest-deep through the snow crust.

Smith, aware of the grumbling among his men, retreated. He left the main party in the green valley below and started out again with a few men across the steep summits of the high Sierras.

Slowly they climbed above the timber line, the horses weak for lack of food, their flanks torn by the jagged rocks, their hoofs slipping on the icy grade. That perilous crossing was later described in plain terms by Smith. To him the moment of glory came after sheer granite walls enclosed the climbers—and suddenly a pass opened, showing light from the east! At last they were sliding, floundering, plunging headlong down the snowy slopes. That was how the first white men crossed the Sierra Nevada.

Ahead of Jed Smith now was the vast, desolate sink of the Great Basin. The next weeks were to reveal to these men the cruelest face of death. As they moved eastward across the plain, the ragged group already knew that survival would depend on water. Smith scanned the terrain for any sign of the elusive river men called the Buenaventura. On countless maps that river was shown running across this basin from the Great Salt Lake to the sea. None of the mapmakers claimed that they had ever washed their feet in its waters. Yet they must have known that the river was here—or did they?

The sun was now white-hot on the burning salt flats. Days passed with no sign of the river. High noon of June 21, their longest day, found Jed Smith and his companions in a desperate fight for life, without water and with little food. They were dragging themselves across the scorching wastes, and each day seemed like their last. And still the Buenaventura eluded them.

Late in the afternoon they crawled under a cedar tree. Smith ordered that holes be dug in the earth. Were these to be their graves? The men scraped in the sand. In these shallow pits they were able to find some cooling relief for their parched bodies. The group slept fitfully that night, tormented by dreams of the mythical Buenaventura and its bubbling waters.

Another morning found the sun blazing high and hot in the heavens. Robert Evans, a member of the party, dropped and lay motionless. He could go no farther. Evans persuaded the others that no purpose would be served if they stayed to die with him. Smith and the rest of the party sadly left him and moved on doggedly toward a distant mountain.

Three miles farther the half-crazed men found a trickling stream and plunged headfirst into the water. Smith's next thought was to get back to their fallen comrade. As he filled a kettle Smith heard two shots in the distance. He hurried back to the place where they had left Evans, fully expecting to find him dead. But the trapper was still gasping for breath as Smith lifted the kettle to his lips. The dying man had fired his gun to guide his companions back to where he lay in a last hope that they might have found water.

Revived, the party moved on. Slowly they inched their way across the salt crust, naked to the waist and half-dead. Three more days of torture brought them to the edge of the Great Salt Lake.

They made their way into the hills to join a large roundup of trappers who gathered here each year for a week of revelry between seasons. And here Jed Smith's party retold their adventures. They had survived the journey. But only Old Smith looked back on it as a success. He had explored new land and had seen the West as no white man had ever seen it before.

It was only weeks before Smith was again on the exploring, mapmaking trail. This time his sights were on the long Pacific seacoast leading to the mouth of the Columbia River, where Lewis and Clark had reached the sea.

More expeditions followed in which Smith combined his probing of new regions with the trapping of beaver. His interest in maps led him deeper into the wilderness, groping and guessing at what lay, unseen, just ahead. Smith's wanderings ranged from the Oregon Trail far to the north to the Santa Fe Trail across the Southwest. He put on his map the natural routes through the West, showing the mountain passes and the navigable rivers, the overland trails and the sheltered valleys. His map was a great breakthrough for truth. And he filled in much of what had been blank. For Smith the myth of the Buenaventura River died along the trail, one of the last falsehoods that had endured on the map of America.

Jed Smith's maps lacked finish and polish. There was a crudeness in his drawing. He was not a trained cartographer, and he never learned the full art of translating nature into the language of lines and colors and symbols on paper. He knew little about the methods which mapmakers use to show the flow of rivers and the shape of shore lines. He did not know the symbols for natural and man-made features. He was not skilled in showing mountains so as to give some idea of their height and shape and the steepness of their slopes. Actually, it would be some years before cartographers would invent numerous methods for showing mountains as a panorama, appearing as though they were seen through the eye of a bird in flight.

However, Smith overcame his shortcomings in his own way. If the map lines did not tell the full story he scrawled in his own notes. Wherever his drafting skills failed him, he described the scene in written words.

Smith's handmade maps disappeared in time and may be lost forever. But before they vanished, they were used by many mapmakers. And the information contained in his maps found its way into the work of every skilled cartographer for decades later.

By 1830 Jed Smith was back in St. Louis. He had achieved the trapper's old dream, making a modest fortune from the sale of furs and escaping from the wilderness with his life. That winter he bought himself a small farm and a house in town, preparing to settle down. By spring, however, the old restless urge had taken hold again. When the desert flowers bloomed in May, Jed Smith was far out along the Santa Fe Trail, finding happiness in what other men would have considered hardship.

However, on May 27 Jed Smith set his last trap and blazed his last trail. In a flurry of violence, the spears of Comanche Indians brought him down to a quick death.

He was the last great mapmaker of his type. From here on, the map of America would pass into the hands of men trained in science and mathematics. They would use fine instruments and master complex skills. But none would forget their debt to Jedediah Smith.

In the spring of 1842 one of America's noted mapmakers lay dying, his work unfinished. He was Joseph Nicollet, a French immigrant who had brought a new quality of skill and science to the mapping of America.

Back in France, bad luck in financial dealings had cost Nicollet his fortune. Penniless, he came off a ship at New Orleans in 1832 with nothing but his knowledge of mathematics and cartography. Somehow he was drawn to the upper Midwest where a French priest with a similar surname had once explored. Like Jean Nicolet two hundred years before him, Joseph Nicollet probed the Great Lakes region, gradually moving westward into the unsettled country between the Mississippi and Missouri Rivers. Fresh from the refinements and comforts of Parisian life, this mapmaker plunged deep into the American wilderness.

In the next ten years Nicollet added something totally new to America's maps. Never before had the landscape been drawn with such accuracy and precision. Nicollet brought into use a new array of mapmaking instruments. He proved his extreme patience at taking the innumerable observations of position, distance, and direction that are required in modern cartography. The mapping of America, previously carried on largely by talented amateurs, was gradually being taken over by scientific experts. And Nicollet was in the forefront of the changeover.

When at last he came out of the wilderness, his strength was

gone and a dream was still to be fulfilled. Nicollet worked tirelessly at a map which he hoped would be a model for future mapmakers. But illness struck him down before he could finish his
project.

To his sickbed in Baltimore came his assistant, John Charles
Frémont. This was a youthful Southerner who had worked at
Nicollet's side in Minnesota and the Dakotas. Frémont brought
his young bride, Jessie, to meet the man who had taught him
everything he knew about cartography.

The stricken mapmaker and the newlyweds chatted that day
about the adventures on the wilderness trail. With laughter Nicollet told the story of how he had once invited Sioux Indian chiefs
to a feast which the Frenchman had prepared himself. The Indians took one taste of the food and threw down their spoons,
certain that they were being poisoned. In the French manner,
Nicollet had flavored the buffalo soup with a substance unfamiliar to the Indians—cheese. It was only with some difficulty, said
Nicollet, that he was able to convince his guests that Frenchmen
often ate cheese—and survived! Before Frémont left that day,
Nicollet had assured himself that this young man would not only
complete his unfinished map but would also carry on his work.
Within three months, Frémont was heading westward with a
mapmaking expedition of his own.

At the age of twenty-nine, Frémont was being entrusted with
an important mission. He was now a lieutenant in the Corps of
Topographical Engineers, under orders to survey the frontier.
Specifically, he was to bring back a precise map of the pioneer
trails across the Rockies.

Frémont was the kind of man who might well be matched
against the Western wilderness and its deadly dangers. He was
brilliant and daring and ready to take on almost any challenge.
It was only later that his headstrong and reckless nature got him
into serious trouble.

Raised in elite surroundings, Frémont learned to enjoy the rough outdoor life, the tang of the morning wind across wavy grass seas, and nights among strange peaks where only the clustering stars were familiar guides. A man of action, he reveled in chasing the buffalo and collecting curious rock and wildlife specimens. He found a deep satisfaction in charting paths through the mountain maze and in giving names to the spectacular but hitherto nameless landmarks of the West.

June found the expedition far out along the Platte River. The journals and diaries which Frémont and his men wrote vividly described the scene. At sundown one day, the company made camp on a high, eroded bluff. Frémont and his two main aides put their heads together over maps. One was Charles Preuss, a nervous, red-faced German, hired as cartographer for the expedition. The other was a frontiersman who had already become a Western legend, Kit Carson.

Frémont could not have picked a better pair, but they were as different as West and East. Carson was no hand with maps. But what he knew of the frontier was etched in his weather-worn face. The scout wore the scars of Indian encounters, and he had learned by hard experience the lore of pathfinding and survival on the trail.

Preuss was unsuited to the frontier life. But he clenched his teeth and overcame endless hardship. His bitter complaints, like his notes and sketches of the terrain, were set down each day in his journal. By the time he completed a series of expeditions with Frémont, Preuss was to produce some outstanding maps of the West, drawn according to scientific principles.

Before breakfast that day Preuss made a few quick sketches of the landscape, taking compass bearings on three peaks and estimating the distances to each. As the crew moved forward, new bearings were taken on these same three pinnacles repeatedly in order to verify the estimates on the sketches.

At a distance Preuss spied several groves of trees and began to put them into his sketch. He looked up again—and suddenly the trees were moving! He wrote in his diary: "My woods, which would have looked nice on the map, turned out to be three immense herds of buffalo."

Frémont fumbled with an instrument that would someday change completely the art of mapmaking. He had secured a newly invented type of camera. However, repeated efforts at photographing the landscape came to nothing. It would be years before the camera was developed into a practical instrument for such use.

After dark Frémont brought out his sextant and telescope to observe the stars and find the party's position. By the light of a guttering candle Frémont set down his findings.

As he had been taught by Nicollet, Frémont used a barometer to find his altitude. This instrument consists of a quantity of mercury enclosed in a glass tube, with the mercury rising and falling according to the weight of the atmosphere. Since air pressure is lower at higher altitudes, the barometer is a means of learning one's elevation above sea level. The instrument is a simple one, but its readings must be corrected according to the temperature. The barometer often proved to be too delicate to be carried on muleback over rugged mountain trails or by canoe through the rapids.

The expedition soon ran into high adventure. In fact, the dashing young leader seemed to be on the lookout for peril in a land where it was not hard to find. Experimenting with the use of an inflated rubber boat, Frémont took a group of five men on a wild trip down a series of waterfalls in the Platte River. As he later described it:

> We cleared rock after rock, and shot past fall after fall, our little boat seeming to play with the cataracts. We became flushed with success and familiar with the danger; and yielding to the excitement of

the occasion, broke forth together into a Canadian boat-song. Singing, or rather shouting, we dashed along; and were, I believe, in the midst of a chorus when the boat struck a concealed rock immediately at the foot of the fall, which whirled her over in an instant.

In the next moment, the five men were fighting the current for their lives. Swirling through the water went supplies and equipment. Frémont managed to clamber ashore and helped the others to safety. Preuss swam to a rock, holding aloft his journal and the chronometer. No lives were lost, but foodstuffs and instruments, blankets and baggage went plunging down the river out of sight.

That first expedition proved to be a useful preparation for the longer ones which were to come. The United States government was now deeply involved in the surveying and mapping of the West, and Frémont was filling an important need. The Topographical Corps asked him for accurate maps that would open the way for westward expansion, maps showing emigrant and military routes, areas for land distribution, and possible paths for transcontinental railroads. The Continental Divide was still the official western boundary of this nation. But the government was eagerly eyeing the vast regions beyond.

In those years, possession of Oregon, California, Texas, and the rest of the Southwest still hung in the balance. However, there were many who believed that these prizes would fall into the hands of the United States—if enough Americans settled in these frontier regions.

The United States government was doing everything it could to encourage the westward migration. The trickle of migrants became a steady stream, and soon it was a full-flowing tide. White-hooded covered-wagon trains were moving slowly across the buffalo plains, drawn by yoked oxen. Land-hungry homesteaders were following a vision of a new life in the West. They were attracted by its spaciousness, the bounty of richness in the earth, and the awesome natural beauty that opened before them.

There were no fixed roadways across the vast plains, and the mountains rose as a barrier that taxed their courage and weakened their hope. But Frémont put into their hands a very useful tool. He produced a map which showed the way from the Mississippi to the Pacific. Drawn by Preuss and printed by the government in seven sheets, this was a large-scale map designed for the use of settlers headed west. It was, in fact, the first great road map of America, drawn on the scale of ten miles to the inch.

Across the plains were detailed the locations of drinking water and grass for the livestock, fuel, and game. There were notes that told about the Indians settled in each area, as well as information about forts and trading posts, fording places across the streams, and easily recognized landmarks. One notice read, "25 miles without water," and another, "First appearance of buffalo."

Most important was Section 4 of the map which showed the route across the Continental Divide by way of the South Pass. With this map each pioneering family held the secrets wrested from the Rockies by Lewis and Clark, Jedediah Smith, and the numberless other explorers who had struggled through these mountains. The Frémont-Preuss map had the added features of greater detail and accuracy.

Along with his maps, Frémont sent back to Washington his reports which described the West in terms that stirred the blood of venturesome Easterners. Printed in large editions by the government, these reports were not the usual dull official documents. Instead, Frémont made them read like Wild West adventure tales. And they sold faster than paperback novels.

The handsome, courageous mapmaker caught the imagination of an America on the move. He was called "The Pathfinder," a nickname that became very useful years later when Frémont ran for President of the United States. Frémont was now something of a hero. His adventures were followed daily in the penny newspapers. At times he was reported lost—only to turn up again with some new triumph of exploration. Thousands thrilled to his stories of action and danger. And they were lured westward by his description of the land and the opportunities.

Frémont helped to shatter a myth that the great plains were a kind of barren desert where nothing would grow. After reading his accounts and studying his maps, farmers flocked toward the unsettled regions which were later to become the "breadbasket of America," wheat-growing Nebraska and the Dakotas. Frémont helped to draw migrant families to the fertile valleys of the Far West, their wagon trains flaunting the banner, "Oregon or Bust!" And he tempted them with his descriptions of the golden beauty of California. Later the discovery of gold would attract eager thousands, and a Frémont-Preuss map would be the first to show the gold fields.

The key to Frémont's success was his ability to surround himself with capable aides. Moreover, the mapmaker had excellent connections in the nation's capital. His wife Jessie, who lived there, edited and livened up his reports, adding literary flavor to them. His father-in-law, Senator Thomas Hart Benton, was a thundering voice in Congress urging forward America's winning of the West.

Wherever Americans gathered in those years, the West was the burning topic. Wherever a map was posted—in a courthouse, a schoolroom, a crossroads store—people studied and argued over the disputed borders.

Four regions drew their attention. Two of them, New Mexico and California, were provinces of the young Mexican nation, which had newly liberated itself from Spanish tyranny. Another region had been a Mexican province but seceded in a bitter revolt and proclaimed itself the independent Republic of Texas. The fourth embattled area was far to the northwest, the Oregon country, claimed by both Great Britain and the United States.

Across America the clamor of the argument grew deafening. Noisiest were those who shouted for American expansion. Some wanted Oregon taken as far as the 49th parallel, which was the existing Canadian boundary. Others demanded that the American border be extended above the 54th parallel to the lower reaches of Alaska. They trumpeted the slogan, "Fifty-four forty [54° 40′] or fight!"

As for California, the American empire builders saw an easy opportunity for taking this Mexican province over completely. Mexico held a weak grip on this sparsely settled region. Franciscan missions were spaced out along the coast like a string of rosary beads, a day's horseback journey apart. Around a few fort towns lived a mixed population with little loyalty to the new government in Mexico City. The large province called New Mexico, including what later became parts of five Southwestern states, was even less populated or fortified.

However, Texas was the explosive issue which threatened to blow up into full-scale war between the United States and Mexico. The government in Mexico City vowed it would fight if the United States made any move to annex the Texas Republic. In Washington, however, the policy was to take Texas—"by purchase, if possible; by war, if necessary."

Through combined military action, diplomacy, and purchase, the United States had been able to acquire Florida from the Spanish. It was argued that the same methods might be used to bring the Western lands under the American flag.

This was a great period for blustering flag wavers on both sides of the border. Appeals to patriotism, red-bloodedness, national honor, and destiny resounded. Many Americans believed the time had come to extend the map of the United States "from sea to shining sea." This brash young nation seemed willing to take on any other country which stood in the way of its growth.

There were, however, voices of doubt. A newly elected congressman, Abraham Lincoln, thought his way slowly to the opinion that war with Mexico was wrong. In Philadelphia protest meetings were held at which speakers opposed grabbing territory from the sister republic of Mexico, which had, like the United States, recently won its freedom by revolution. In Massachusetts the poet and naturalist, Henry David Thoreau, went to jail rather than pay taxes in support of the Mexican War.

During all this tumult, John Charles Frémont was moving into California with a mapping expedition. The Mexican authorities looked on these visitors with mistrust. Frémont's company was too large for simply mapmaking, and heavily armed. In the weeks that followed, the officials were to learn how well-founded were their suspicions. The bold young Frémont turned his mapping party into a military striking force. He gave support to a rebellious band of American settlers who proclaimed California's independence from Mexico and unfurled a flag displaying a crude drawing of a grizzly bear.

Frémont marched his little army up and down the coast. He had now completely given up the pretense of mapmaking and was openly involved in the military and political battles which were being fought across the map of America. However, the impetuous Frémont soon rushed into military ventures that might have proved highly dangerous for the United States. He defied his superior officers and carried out a series of high-handed actions without authority.

By the time the Mexican War was over, Frémont's fortunes took a downward turn. Because of some of his foolhardy actions in California he was court-martialed, found guilty of disobedience, and ousted from the United States Army. Shamed and angered by the verdict, Frémont immediately sought some new way of proving himself and humiliating his enemies. He vowed to carry out some stunning feat that would bring him back to glory and even greater fame.

In Washington Frémont was consulted by a group of men who were considering the building of a railroad from St. Louis to San Francisco. They drew a line on the map running along the 38th parallel. Could such a route be found across the Great Plains and the Rockies? And could a railroad be built along that line, maintained and run year-round, winter and summer? Frémont was sure he could map such a route. What was more, he would answer their question by crossing the mountains in midwinter!

Frémont's expedition started out in the soft haze of Indian summer, 1848. His party numbered thirty-three men, among them many of his old stand-bys, including the cartographer Preuss.

By mid-November the expedition was in serious trouble. Severe storms had blanketed the mountains in deep snow. Against the advice of Indian guides and veteran mountaineers, Frémont ordered his company up the slopes.

The weather worsened. In the next weeks the men fought their way up craggy peaks in a disheartening struggle for every foot, every inch of icy ground. Shrieking gales swept across the mountains, hurling sleet at the climbers. Sleep was impossible. The high wind blew out the campfires. Every trace of animal life had been driven from the mountains—except for this band of half-frozen men and their dwindling herd of pack mules.

The stubborn Frémont would not stop or turn back, nor would he change the disastrous course he had planned. Each morning found more of the mules frozen dead. Supplies and rations had to be discarded to lighten the loads on the remaining animals. The endurance of the men was giving out. And now Frémont began to realize that he was hopelessly lost.

On Christmas Day, savage new storms struck. Frémont began his retreat, aware that his whole party faced death by freezing or starvation. By the time that expedition ended, eleven men had died. The rest staggered snow-blind and wild-eyed, frostbitten and starving, out of the mountains. Frémont was forced to admit defeat.

In the next years the career of John Charles Frémont went up and down—while the fortunes of America changed as well. The mapmaker who had been disgraced in California returned there to become the state's first senator. Once drummed out of the United States Army, Frémont became a general in the Civil War. He was nominated for the presidency of the United States but failed to win election. Frémont struck it rich in California gold, only to end his life penniless and in debt.

The mapmaker was a colorful figure on the American scene all through the years when America's map was changing. By the 1860s the main land area of the United States was basically what it is today. From defeated Mexico, all of Texas was acquired down to the Rio Grande, as well as the New Mexico territory and California. Oregon was also firmly in United States hands. A section of land along the southern border of Arizona was secured from Mexico through the Gadsden Purchase.

In the year 1862, the map showed thirty-four states in the union and eight territories. The nation occupied one continuous land area between the oceans, held together not only by its stormy history but by its unifying geography as well.

And yet a deep-burning struggle between North and South threatened to tear it apart. The bitter battles of the Civil War were being fought that year across the line which had been surveyed a century before by Mason and Dixon. But a sober and confident President Lincoln was certain that there was no line along which the compact nation would split itself into two separate nations. He invited all Americans to see the oneness of the country "as shown by a glance at the map.

"Trace through from east to west, upon the line between the free and slave country, and we shall find that a little more than one-third of its length are rivers, easy to be crossed . . . while all of its remaining length are mere surveyor's lines," Lincoln declared. "There is no line straight or crooked, suitable for a national boundary upon which to divide."

From towering Fremont Peak in western Wyoming, the melt trickles down toward the Green River. And where the rivulets gather to form a broad stream, a bit of double drama was enacted in the spring of 1869.

On a bright May morning a tiny wood-burning locomotive of the Union Pacific clattered across the high trestle. The train was making the historic first run westward across the United States.

From the windows of the two cars the first transcontinental passengers stared down into the river below. There a group of men were getting ready to shove off on a momentous mission of their own. They planned what no one had achieved before— the exploration and mapping of the raging Colorado River.

Head of the expedition was John Wesley Powell, a short stump of a man who grew a bushy set of whiskers when he was on the trail. For him the river had become a living challenge. Its roaring flow resounded in his dreams. And in his waking hours his mind grappled with the problem of navigating its length.

Like all other dark corners of America before they were explored, the Colorado canyons were veiled in fears and fables. The river was a whispered terror in mountain lore and in Indian legend. No man was a match for its violence. There were places, it was said, where the devil river suddenly plunged beneath its rock walls, sucking its victims down into a watery vault.

John Wesley Powell was no believer in demons. But he carefully checked out the accounts of explorers who had attempted to run the Colorado and failed. Along the way he was to find the grim relics of their agony.

One of America's longest rivers, the Colorado begins in the high ridges of the Continental Divide. Gathering force and speed, it thunders down from the steep heights on its way south and west. It drops two and a half miles before it finally reaches the Gulf of California. The river descends in great leaps through high falls and foaming rapids.

For centuries the Colorado has pounded its burden of silt against its stone beds, slashing a groove into the broad plateau through which it runs. Its masterwork is the Grand Canyon, ten miles wide and a mile deep. Pent up between its sheer cliffs, the Colorado becomes fuming white-water fury. It pitches and plunges like the open sea in a ceaseless storm. And this was the river that Powell chose to ride—in a rowboat!

Powell had long since tested his courage on the frontier where he grew up and received a homespun education. Young Wes went to a one-room schoolhouse, sitting on a split-log bench and figuring sums on a slate. By the age of eighteen he was a teacher in a similar schoolhouse.

As his pupils moved up the educational ladder, Powell moved too. By 1860 he was teaching natural history in an Illinois college and exploring the wilderness in his spare time.

As the sound of the Civil War's guns grew louder, Powell said good-by to a classroom which was cluttered with his collection of shells and rocks, fossils and Indian relics. An outspoken foe of slavery, he answered President Lincoln's first call for volunteers. Wes Powell came home with the rank of major. Missing was part of his right arm, blown off in the battle of Shiloh.

The frontier had moved west, and Powell moved with it. In the plateau and canyon country he found riddles enough to challenge his searching mind and his restless energy. He saw these regions as newly opened lands whose natural resources were already being threatened by spoilage and waste.

Men of science in those years were beginning to piece together a deeper understanding of the earth and its land forms. They were finding orderly patterns to explain uplands and drainage, the surface shape of the land and the rock layers beneath. Their studies revealed an ever-changing planet, its crust gradually re-molded again and again in the course of countless centuries. Repeatedly the lands were invaded by the encircling seas. Moving glaciers covered great areas and disappeared, leaving behind layers of mud and rock. Mountains slowly heaved upward. The land forms were constantly being changed by the grinding action of water and wind.

Powell became eagerly interested in the search for the truth of how the land took shape. He began to think of the Colorado River as a matchless laboratory of earth science. In gouging through the rock, the river had exposed to view a record of the changing land going back to the most ancient of ages. The full account was spread out on the walls of the canyons, the layers of change clearly marked off in variations of brilliant color.

The little Major knew the perils of this river. But if he could explore it and map it, what a story that map would tell!

It was a pitifully small and unimpressive party that launched four frail little boats down the Green River. With no financial backing from the government or any large institution, the expedition was rather poorly equipped and manned. Powell had nine men, of whom he was by far the best educated and experienced.

Where the swift-running Green River entered the Colorado mainstream, one man departed, declaring that he had "seen danger enough." The boat in which he rode had been smashed to pieces on rocks, and he had barely escaped with his life. The wreck cost the company dearly in supplies and equipment.

Powell put the task of drawing the map into the hands of one of his men, a printer named Oramel G. Howland and called "O.G." Untrained as a cartographer, still Howland did a creditable job, working according to a simple routine. He took a compass reading at each bend in the river, then reckoned roughly the distance between bends. In fact, the men in each boat made distance estimates of their own, and all these were averaged out.

Howland's map sketches were laid over a crisscross grid showing the longitude and latitude, which was checked by astronomical observations every fifty miles.

Altitude was determined by means of the barometer. Readings were taken along the river shore, and wherever possible the canyon cliffs were climbed and their height measured. Powell took charge of exploring the canyon formations, measuring the width of the exposed earth layers, taking samples of the rocks, making notes on the region's natural history.

Usually it was the Major and George Bradley who scrambled up the sheer walls. Despite his handicap, the Major was tough and an agile climber. He and Bradley kept passing the delicate barometer back and forth between them as they made their way up the cliffsides.

At one place, however, Powell became so absorbed in studying the rocks that he suddenly found himself hanging from a rocky ledge by his one hand. Below him was nothing but the dizzying drop into the raging river. His outcry brought Bradley to the rock shelf above, just out of reach. Without a word Bradley quickly stripped to the skin. He carefully lowered his long underwear toward the trapped Powell. With one motion Powell let go of the rock and grabbed the dangling drawers, and Bradley slowly pulled him up to safety.

As his surveying and mapping party made their way down the Colorado, Powell left behind markers recording their observations. He was part of the great effort to map America's West. And Powell had chosen for his share the last completely unknown section of the frontier and the most difficult to explore.

Other surveying parties were working in the mountains to the north and in the newly formed states of the Southwest. From the Pacific seacoast a government-sponsored survey headed by Clarence King was exploring eastward along the 40th parallel.

The large expeditions were using new methods and were

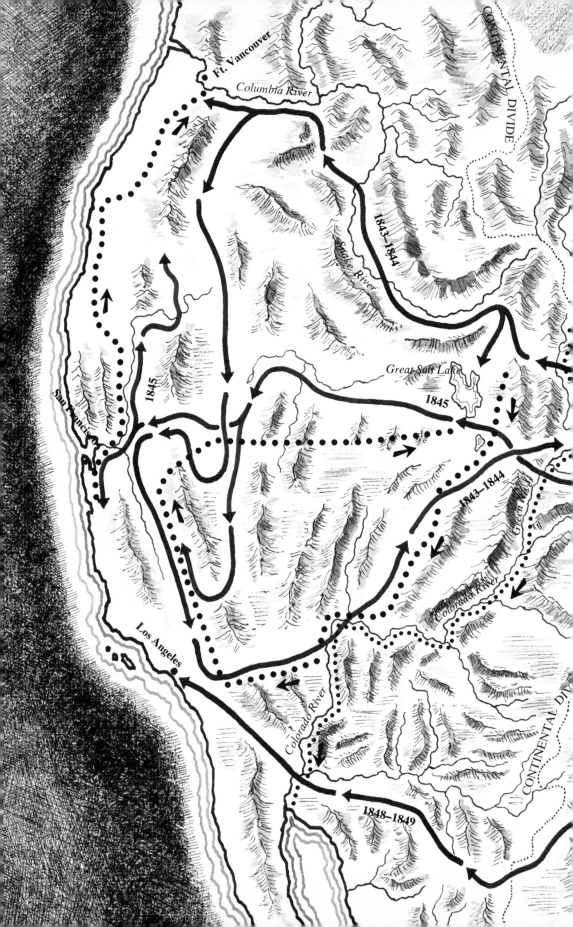

EXPLORERS OF THE WEST

JOHN C. FREMONT ━━━━━
JEDEDIAH S. SMITH ●●●●●●●
JOHN WESLEY POWELL ●●●●●●

Missouri River

Yellowstone River

...ont Peak

Ft. Laramie

1842

Omaha

John Wesley Powell

Platte River

1843–1844

Missouri River

Mississippi River

Pueblo

Kansas City
(Independence)

1845

St. Louis

Arkansas River

...48–1849

Pecos River

Jedediah Smith

John C. Fremont

equipped with improved instruments. Longitude was determined by the time signals received by telegraph from observatories. Triangulation over great areas was achieved by catching light signals from distant peaks or sighting on towering rocks or lone trees, measuring the angles between stations observed. The camera was being used with more success. As they completed their observations, the surveying parties were leaving their bench marks fixed into the ground, inscribed with survey information.

In pure fun Powell's men held a small ceremony one day and implanted a "bench mark." It was one of the tough little biscuits baked by the Major during his turn as camp cook!

However, lighthearted moments grew fewer as the danger and the drudgery of the journey increased. The rockbound river made toys of the lightweight boats, splintering the oars, spinning the craft in whirlpools, ripping open seams, and flinging the party headlong down chutes and cascades.

Powell tried to estimate the risks ahead by examining the nature of the rocks. Soft rock usually meant quieter waters; hard rock warned of a river running rampant among jutting boulders and seething rapids. In the most troublesome spots the boats were slowly passed along the water's edge by means of ropes held by the men as they clung to the canyon walls. However, in order to move faster, the Major sometimes took a chance on an unknown hazard. Occasionally he guessed wrong, and the boats capsized, the men trying to save themselves and whatever cargo they could salvage.

Throughout the trip the Major kept a daily journal. To read this account is to move, here and now, with the expedition into the savagery of the stream and the dread of the enclosing canyon. All was set down just as it happened. Powell's language is terse. But sometimes he was carried away by the grandeur of the scene.

Powell and Bradley worked their way painfully three thousand feet up the cliffside. "At its top, what a view. Its walls are set with crags and peaks, and buttressed towers, and overhanging domes," Powell recorded. "We look up Whirlpool Canyon, a deep gorge with a river in the bottom—a gloomy chasm, where mad waves roar. But at this distance and altitude, the river is but a rippling brook, and the chasm a narrow cleft. The top of the mountain on which we stand is a broad, grassy table, and a herd of deer is feeding in the distance. . . ."

Powell drew on his imagination to give names to the landmarks which would appear on maps for all time. He called them Split Mountain Canyon, Flaming Gorge, Sockdolager Rapid, Vermilion Cliffs.

In some of these canyons, the colorful sandstone walls were pocked with caves. A series of rain-water pools formed a staircase. The canyon walls were bathed by gushing springs and lacy waterfalls. A single gunshot rattled like drum beats, echoing from cliff to cliff to cliff.

In his journal Powell was also an instructor, explaining the meaning of the landscape. "You must not think of a mountain range as a line of peaks standing on a plain," he wrote, "but as a broad platform many miles wide, from which mountains have been carved by the waters."

To Powell the Colorado was a catch basin of the region's bygone days. "Past these towering monuments we glide," he put down in his journal, "until we reach a point which is historic." The party found the ancient cliff dwellings of Indians who had once lived here and had left behind bits of their pottery, picture-writing in stone, and tools.

In the sixteenth century Coronado's men blundered to the edge of the Grand Canyon. They gasped at what opened up below them, and they shuddered at the thought that another step would have taken them over the yawning brink of the earth.

One hundred years before Powell the Spanish priest Escalante, accompanied by the mapmaker Miera, came up against the barrier of the Colorado. That overland party edged the canyon until they found a place where they cut steps into the rock, brought their horses down and forded the river. The location later appeared on maps as the Crossing of the Fathers. Powell's expedition would honor this courageous leader when they found a broad stream which had never appeared on any map. This, the last undiscovered river in America, was named the Escalante.

But Powell's interest was centered on an older past, older than man. "All about me are interesting geological records," he wrote; "the book is open, and I can read as I run."

The men of the expedition were now overcome by fatigue, tired of being exposed to sunburn by day and freezing cold at night. They were fed up with spoiled bacon and wet flour, soaked blankets and leaky boats. The younger men in the crew showed signs of homesickness and a yearning for home cooking. One day the Major noticed one of the men fumbling with a sextant and asked him what he was doing. "I was just trying to find the latitude and longitude," came the reply, "of the nearest piece of apple pie!"

The constant danger was beginning to prey on the men's minds. They slept uneasily in dread of the ferocious rapids just ahead. One night, when the party had reached a place where the canyon wall was an easy climb, O.G. Howland came to the Major with a troubled face. He proposed that the remainder of the river trip be abandoned, declaring that it was madness and suicide to go on.

That night Powell paced the river bank, waking his men one by one for a brief chat. By morning he had made his decision. He would continue to ride this river until, like some wild stallion, it ran out of fury.

Howland and two others decided that they valued their lives too much to go on. After a sad parting they disappeared over the

rim of the canyon, headed for settlements seventy-five miles distant. Weeks later the party learned the fate of Howland and his
two companions. In a dispute with Indians all three had been
killed.

Almost four months after it had begun, the Colorado expedition reached its end. It was August 30 when the exhausted little
band came upon two men and two boys who were dragging nets
through the river. These settlers were looking for the remains of
the Powell party which, according to reports, had perished in the
Colorado River.

But the six men were very much alive. They were the conquerors of the demon river. Uncovering the record of its past,
they had made some history of their own. Their voyage had un-
sealed the last great unknown on the nation's map.

For the crew the journeying was over. But for Powell this was
only a beginning. His concern about the heedless way the West
was being settled carried him back into the Western wilderness
again and again—and at last into Washington.

In the nation's capital Powell found challenges tougher than
the Colorado. His battle now was to help bring about a sensible
public policy for mapping, exploring, dividing, settling, and conserving the Western land and its natural resources. With anger
and dismay, the Major told a shocking story to the lawmakers.
What he had witnessed in the West frightened him. There were
dangers to the country greater than the menace of the Indians.
In fact, the well-being of lands which the Indians had long pre-
served was in peril. Railroad, mining, lumber, and land-owning
interests were threatening the resources of the frontier.

At that moment, he pointed out, schemes were under way for
running a railroad across what is now Yellowstone National
Park and for pockmarking its terrain with numerous lead and
copper mines.

Using his map as a weapon, Powell fought for better land use policies and for the protection of the public domain against the private interests which were overrunning the West. He entered the government service and began a determined battle to save precious water resources.

West of the 100th meridian, the Major explained, was a vast land with little rainfall. There the small farmer who could not reach water was doomed. Powell had seen how the limited water supplies of those areas were falling into the grip of the wealthy interests. The Major argued that the system of surveying land which portioned out small 160-acre family farms in Ohio and Wisconsin was not suitable for the arid West. He deplored also the setting up of state and county borders according to surveyor's compass lines, instead of tailoring them for a fair division of drainage. "There will never be enough water," he declared, pleading for irrigation systems, the creation of reservoirs, and the formation of cooperative farm groups which could deal fairly with the water problem.

Although fearlessly outspoken, Powell was no match for the powerful interests which were out to wrest great wealth from America's public lands. Long afterward it was said that if the little Major had been heeded, the Western states might have been spared the spoilage of the land, the pollution of the streams, the disaster of the dust bowls, and tragedies of drought and flood.

In those years, it was just becoming clear that the federal government would have to perform scientific research that no private organization was willing or able to undertake. A self-taught scientist, Powell continued a lifelong program of educating himself. He also fought for the application of science to many public problems. The schoolteacher Powell was still a teacher. He taught his co-workers what he knew, turned his projects over to them, and went on into new fields.

By the 1870s the mapping and surveying of the nation had become ensnarled in bitter rivalries. A dozen different government agencies were in the field, their work often overlapping, duplicated, and conflicting. Sometimes crews clashed on the trail.

"Powell of the Colorado," as he was still called, was seen as the man who could unify the nation's map work into a single efficient effort. Restless, energetic, and well organized, Powell plunged into the difficult task. Under his new plan, the United States Coast and Geodetic Survey was asked to continue its excellent work of mapping the nation's shore lines. The Corps of Engineers, whose exploratory work dated from the Lewis and Clark Expedition, remained in charge of military mapping. However, the major map work was unified under the new U.S. Geological Survey. In 1880 Powell became its director.

The Major was now ready to advance his boldest project—a national map of the United States. It was to be a far-reaching, costly, long-range undertaking, but the need for it was obvious. When Powell went to the Capitol to present his plan, congressmen were appalled that no such map had ever been drawn.

"Then we have no official map of the United States defining its frontiers?" one legislator asked him.

"No, sir."

"And we have no official map showing the political boundaries within the nation?"

"No, sir, not with any degree of accuracy."

Under Powell's direction the national mapping job would be carried on piecemeal, with hundreds of surveying parties working at once throughout the country. The final result would be a series of maps, small enough to be easily handled and inexpensive, each an accurate representation of a small section of the United States. Together these so-called "quadrangle maps" would present the complete likeness of this nation.

Powell aimed at a type of general-purpose map that would also serve a great many specialized needs. He would not be satisfied with anything less than a high level of accuracy and the kind of clear form that would bring its use well within the scope of the average citizen untrained in map skills.

With his staff the Major worked tirelessly at standardizing a design, a set of elementary symbols, color schemes, lettering, and a method of indicating place names. Today the quadrangle maps, greatly improved in accuracy and appearance, are published very much in the style that Powell envisioned. These maps are fixed in location by a grid system, bounded by parallels and meridians.

Numerous methods had been devised by cartographers to convey the idea of the height of the land forms. Powell insisted on the use of contour lines. These are flowing lines of irregular shape which connect points at the same elevation above sea level. They are like the lines often formed around the inside of a water barrel as the water level falls. Contour lines clearly reveal the shape of the relief features on the land. When they are irregular and zig-zag, they denote craggy, rough terrain. When contour lines are close together, they depict a steep slope. These lines are printed in brown on the quadrangle maps. Each fifth line is a slightly heavier one and is labeled with the altitude in numerals.

Water features are shown in blue, wooded areas in green, and man-made objects in black. Road classifications and public land surveys appear in red. Uniform symbols, used throughout the world, were selected for these maps with the idea of making them easy to understand. A marsh is represented by a simplified design which looks very much like a marsh. The orchard symbol shows regular rows of green trees. Brown dots, varying in density and pattern, represent different types of sand and gravel surfaces.

Man-made features, such as bridges, dams, tunnels, and canals, appear like these objects as seen from above. A railroad is a long line with crossties. A school is a small square topped by a flag. Quarries and gravel pits are denoted by crossed tools.

The selection of a uniform scale for the national map project proved to be a knotty problem. Vast desert areas might be adequately handled by means of small-scale maps. However, the built-up communities, where there is much variety and activity, obviously needed a larger scale in order to show fine details. Depending on the type of region, quadrangle maps are issued today in several scales. Three of the most common are: one inch represents 2000 feet (1:24,000); one inch represents nearly a mile (1:62,500); one inch represents nearly four miles (1:250,000).

In this growing, expanding, changing nation, maps become obsolete fast and must be constantly brought up to date. Perhaps Powell did not fully foresee that maps would sometimes be out of date almost before they were off the presses.

Powell believed the national map could be finished in twenty or thirty years. However, it is now hoped that the detailed mapping of the United States will be completed by the year 1981—a full century after John Wesley Powell began it.

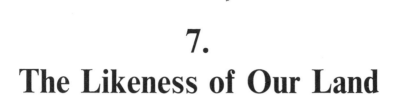

7.
The Likeness of Our Land

Our map is an eagle's-eye view of America. Seen from on high, the panorama spreads out across a vast land area and beyond the sea. The country is richly varied—and changing. Nature is at work remolding the earth forms. Man is also making his mark on the landscape.

Study the map and it will reveal not only America as it is today but also traces of what it used to be. The map tells how the time-less streams cut their paths through the earth. Disclosed also are patterns of men moving restlessly across America, from the coasts deep inland, from the seaboard settlements into the back

country, and more recently from the countryside toward the urban areas.

The map says, "This is how it was"; and also, "This is how it is."

Powell's Colorado River is no longer quite the same. At one point where the mapmaker fought rock-strewn rapids the river trickles today through the spillway of a mighty dam. And where torrents once boiled in a furious white froth, there is now a deep blue lake, a gleaming mirror reflecting the red sandstone cliffs. This is newborn Lake Powell, a vast reservoir which has suddenly appeared on America's map in recent years. Perhaps nothing could better emphasize that the map is a changing thing, forever incomplete.

Today mapmakers work hard at keeping up with changes that nature, too, makes in the landscape. The Maine coast, mapped by Champlain, is being inched away by the sea, leaving only the hardiest rock to resist the pounding waves.

The Mississippi River has changed course a thousand times since Jolliet mapped it. The "Father of Waters" continues to meander into new channels and short-cut across its horseshoe bends.

A famous Mississippi boatman and storyteller, Mark Twain, wrote about this wandering river and how it moved in close to towns that were previously without a waterfront and left former river towns high and dry. The Mississippi village of Delta, once downstream from Vicksburg, was upstream from Vicksburg after the river cut across a narrow neck of land and found a new path. "A cut-off plays havoc with boundary lines," mused Twain. "Such a thing happening on the upper river in olden times could have transferred a slave from Missouri to Illinois and made a free man of him."

At its mouth the Mississippi River has created a vast delta ten times as big as the state of Rhode Island. The Gulf of Mexico and the ocean shore lines are being constantly altered. Moving waters carry sand and mud from one place and deposit them in another. Cape Kennedy, where spacemen take off for new worlds, is itself new land, built up by the sea from a sand bar. Thousands of New Jersey's acres have been carried away by the waves, while an almost equal number of acres of beach have been formed elsewhere. The well-known New York City recreation area of Coney Island is a gift from the sea.

Wind action shifts towering mounds of sand at the Great Sand Dunes National Monument in Colorado and at the White Sands National Monument in New Mexico. Water actions join in this process to move dunes along Cape Cod, Massachusetts, and at the lower end of Lake Michigan.

Great glaciers changed America's landscape in the past. But there are still live glaciers at work in Montana, Washington, and Alaska. Parts of the Alaskan shore line are not land but glacial ice. Since these glaciers advance and recede constantly, the maps quickly become obsolete.

Maps are quickly outdated by violent natural forces that suddenly rip the terrain apart, heave up bedrock formed when ancient seas receded, spout hot lava high into the air, and cover the landscape with volcanic debris. The maps of the Rockies, the Pacific Northwest, the Southwest, and Alaska show many old volcanoes. Hawaii has active ones that erupt periodically.

On an August morning in 1959 a surveyor in Yellowstone National Park watched a tempest that suddenly boiled up in his coffee cup. By the time the earth stopped quaking, the top of a mountain had rolled down and dammed up the Madison River. Months later a Yellowstone survey crew recorded new land forms, traced a new lake made by the earth dam and called it Earthquake Lake.

On the early maps of America, the works of man appeared in

the form of towns and the roads running between them. In time, however, the man-made features have become more and more apparent on the map.

The towns and their names, locations, sizes, and rates of growth tell a great deal about America. There is no other country on earth that has so many town names that begin with the word *New*. Immigrant groups who came here often found some site that reminded them of their homeland, settled, and named their settlement after the place from which they had come.

The maps show cities such as Oklahoma City, Oklahoma; Los Angeles, California; and Houston, Texas, as covering large tracts of land. Many cities are surrounded by numerous suburbs. The cluster of growing cities and towns in some regions of America is rapidly leading to long stretches of built-up urban area, running continuously for hundreds of miles.

This nation also has its run-down towns, ruined towns, and ghost towns. Some once-prosperous communities have disappeared from the map. There are river ports without any river traffic, mining camps without minerals, crossroads towns with grass growing in the streets. Trading posts that once bustled with the business of trappers and frontiersmen, prospectors and pioneers have vanished along with the swarming beaver and the buffalo herds.

America's towns, like those of other countries, are found mostly at the water's edge. Men have made the greatest changes along the watercourses. The maps show where they have dammed streams, drained swamps, filled in new land, dug canals, created lakes, and dredged out harbors.

On so-called political maps, the most prominent and the most changeable features are the political boundaries. Maps of the United States had to be revised in 1959, after Alaska and Hawaii were admitted to statehood. The man who first put both Hawaii and Alaska on the map was Captain James Cook, the great English explorer. Although the Russians got to Alaska long before him, Cook did the first important mapping survey of the coastal regions. In 1778 Cook discovered what he called the Sandwich Islands, later Hawaii. As Cook was mapping these places, the new nation was emerging of which they would someday be a part.

Old maps of the United States tell a story of shifting state boundaries, the changes often accompanied by bitter conflicts. An error by a mapmaker touched off what has become known as the "Toledo War." According to an official map Ohio was deprived of what she considered to be the northern strip of her territory, including the town of Toledo. The governor of Ohio threatened to take back the land by force. In reply the governor of Michigan vowed he would "fire on the first man who crossed the boundary line." A Michigan balladeer of those years told the fearful story:

> In eighteen hundred thirty-five there was a dreadful strife
> Between Ohio and this state; they talked of taking life;
> Ohio claimed Toledo, and so did Michigan;
> They both declared they'd have it, with its adjoining land.

Although strong words were shouted and firearms were brandished, there was no bloodshed. The federal government stepped in and made peace. Ohio was given the Toledo strip. Michigan's pride was also soothed with an award of land. That was how the so-called Upper Peninsula, bordering on Lake Superior, became a part of Michigan. During that tumultuous period, the maps of the region seemed to be changing from day to day.

The border battles between the states continued all through the nineteenth century and into the twentieth. Almost every state has been involved in some such border fracas. Probably the only exception is Hawaii—which continues, however, to struggle for its land against the surrounding seas.

Many of the newer states are framed by meridians and parallels, charted by compass and transit, fixed by the stars or the sun. The 37th parallel now forms part of the boundaries of six states. When Colorado achieved statehood in 1876 and Wyoming in 1890, both appeared on the map as perfect rectangles, their margins following the four main compass points.

Irregular boundaries of the Western states run along the major rivers, including the Columbia, the Missouri, the Rio Grande, and the Colorado. Part of the Idaho-Montana state line follows along the Continental Divide.

Clearly mapped boundary lines have helped to make good neighbors of Canada and Mexico. The United States has long been proud of the fact that it has "undefended" borders. Many times in the past, however, conflicts broke out as both sides claimed the same piece of ground and offered maps as proof.

Where the United States reaches out to its easternmost point, a long-drawn-out fight was carried on over the border between Maine and New Brunswick. The argument went back to the time of Samuel Champlain. An extended debate raged over what was in the French explorer's mind when he named the area St. Croix. Three rivers come together there, forming a cross. And the place named which Champlain put on his map was St. Croix, meaning Holy Cross. In the 1783 treaty between Britain and the United States, the St. Croix River was set as the international boundary line. But which of the three rivers was Champlain's St. Croix?

The argument continued as investigators combed the region carrying copies of Champlain's original map. Their search became a treasure hunt as crews dug into the ground to try to find the remains of where the Champlain party had once lived and the location of the landmarks shown on his map. It took almost fifty years for this dispute to be settled. It is often a simple matter for treaty makers to agree on a line—but another to locate that boundary on the ground.

Another treaty set a line which was bound to cause trouble when the United States purchased Alaska from the Russians. According to that agreement, the border between Alaska and Canada was agreed as running along the peaks of the coastal range —mountains no man had ever climbed!

It was a pig that set off a raging wrangle in the broad strait of Juan de Fuca, where the northwestern corner of Washington state borders British Columbia. The exact location of the channel of the strait, which separates the two countries, had never been clearly established. This left a situation in which a number of islands were not definitely known to be either American or Canadian.

In 1859, on an island in the strait, a pig which belonged to a British official was caught in the act of uprooting the potatoes in the garden of an American citizen. The American shot the pig. This led to a scuffle which landed both men in court. However, the next question that arose was whether the case was properly to be tried in an American or a British court. Now the matter of which country owned the islands could no longer be dodged.

In the next months, the dispute became more and more heated. British and U.S. gunboats faced each other menacingly in the strait. Was a pig to become the cause of war between two great nations?

Fortunately the hotheads were cooled off. Surveyors were hired to locate the true channel of the strait. New maps were drawn. Leaders of other nations were brought in to help render a decision. It was the Emperor of Germany who finally settled the case of the pig!

At the close of the Mexican War the longest part of the boundary between the United States and Mexico was established as the Rio Grande. Unfortunately, this river is a mapmaker's nightmare. Like some mischievous prankster, the Rio Grande frequently changes course, shifting pieces of land from one jurisdiction to the other. The whim and frolics of this river have more than once caused national tempers to flare.

On December 13, 1968, a dynamite blast echoed along the border near the point where the Rio Grande passes from New Mexico to old Mexico. A boundary dispute was being settled— but this time not by gunfire and bloodshed. The explosion opened a new man-made concrete-lined channel. And the ceremony of the day also marked the official return of some 637 Mexican acres which had been transferred by the river to the American side.

The roguish Rio Grande had been put back in its place on the map—for the moment at least. But the restless earth will not sit still while the cartographer draws its portrait. And no map is everlasting.

What is the life span of a map?

A topographical map indicating relief features may be used for many years. However, a weather map, showing temperature, moisture, and shifting winds, lives only for a day.

Of course, even old and obsolete maps are interesting as history, and the National Archives in Washington contain a rich treasury of our mapping efforts through the centuries. But America's mapmaker works hard at keeping up with new changes and new knowledge. From his drawing board comes an up-to-the-minute picture of this nation or a portion of it. From the printing presses rolls an endless supply of fresh maps to fulfill a variety of needs.

Does some city need a new airport? First it will call upon the mapmaker. Is some drilling company searching for new oil fields? Their geologists will be working with maps. Are rampaging rivers compelling steps toward flood control? The area will first be carefully mapped. Is some business firm planning a widespread sales effort? Hardly any such campaign gets under way these days until a map has been stuck full of pins.

Hundreds of private firms produce special maps for a wide range of specific uses. Some of these maps are found in classrooms. Many are in business offices where men are guiding the flow of goods and services. Still others are used by urban and regional planners who carefully observe the movements of people.

The federal government is America's most active mapmaker. The U.S. Geological Survey alone distributes more than seven million maps a year. Most of these are the quadrangle maps, used by sportsmen, farmers, business firms seeking new industrial sites, and government officials working at conserving natural resources.

Dozens of other agencies in Washington, D.C., produce maps which are available to the public, often free or at a small cost. Federal Power Commission maps show where the power lines run. The International Boundary Commissions trace the nation's borders. From the Library of Congress come historical maps showing explorers' routes and battle sites. Many maps that have military purposes are issued by the Corps of Engineers. The Weather Bureau comes out with new maps every day. There are even maps which show how much of America has been mapped!

Gradually the mapmaker is doing much more than charting the land mass of this country, its uplands and valleys, its plains and waterways. His job now often has to do with people. The map has become an important tool in dealing with the problems of man in his relationship to his environment. Today special-purpose maps are being drawn which tell the grim story of polluted air and water—so that we might begin to overcome these problems.

Maps keep a record of our use of natural resources, so that we can be warned of those places where we are dangerously depleting the supplies of minerals, timber, water, or fertile soil. Maps also keep track of how the land is being used in order not to repeat the tragic mistakes of the past.

Today we are trying to think about the future needs of America. And maps are a graphic form of describing the growth and movement of the American people. The mapmaker is called upon to show trends in population, to indicate what the future demand will be for housing, food, roads, public transportation. The cartographer is asked to answer in maps questions like these: How does a city grow? Where are new industries locating? How much of our land needs to be used for food production? Where will increased supplies of electric power be needed in the next decade?

Among the most useful maps are those based on census figures. They reveal where America lives, how we support ourselves, how we are distributed by age, national origin, and income. Still another type of map shows political activities—how America goes to the polls and votes by Congressional districts.

Many maps include the time zones. The noonday sun appears directly overhead one hour later in St. Louis than it does in Baltimore. This was no problem in the years when Americans lived all together in the eastern United States. However, a standard system of time zones became necessary as the nation expanded, and especially as railroads began spanning the continent. Today's maps show four different time belts from Maine to California, each an hour apart, and the Hawaiian Islands and Alaska spread out over three additional time zones.

In the northwestern corner of South Dakota there is a point which is agreed to be the geographical center of the continental United States. The place can be found on the map near the town of Castle Rock, South Dakota.

Restless men, filled with daring and curiosity, have mapped this nation. Today maps are used to guide the wanderings of millions of mobile Americans. Most widely used of all are road maps, the kind that can be picked up free from a rack at a filling station. These are printed in millions of copies each year. They are revised constantly to keep up with the fast-moving road builders.

America's oldest roads were the Indian trails which followed the beaten paths of deer and buffalo. The explorers and map-makers used these same trails. In time they became the well-traveled routes of fur traders and gold miners, pioneer farmers and town builders.

Marked, widened and improved, these roads crossed the wilderness and opened America's deep interior. The primitive roads swerved for every stump and boulder. They wound through the hills in corkscrew turns, often doubling the distance that a crow might have flown between two points. They were dusty in dry weather, muddy in the rainy season, and often impassable in snowtime. But they served a people on the move. The so-called Wilderness Road led early frontiersmen through the Cumberland Gap at the point where the states of Virginia, Tennessee, and Kentucky now join.

Back in 1792 Pennsylvania built the first of the great hard-surfaced roads. This was a toll road, charging a penny a mile for a one-horse vehicle. On payment of the toll a pike-studded barrier was turned around on its pivot and the road was opened. From this arrangement came the word turnpike.

The Spaniards were road builders who linked their forts and missions with sections of what they called El Camino Real, the Royal Highway. Portions of this highway network led to St. Augustine, Florida, to Santa Fe, New Mexico, and to San Diego, California.

Through the middle of America the pioneer traffic moved over what were shown on maps as "traces"—Zane's Trace, the Natchez Trace, the Vincennes Trace, and many others. Often these were improved roads, their beds lined with logs or planks.

However, the furrows and ruts, potholes and puddles were often enough to discourage travelers or to stop them completely in their tracks.

The main corridors across the West were the Santa Fe Trail, the Oregon Trail, and the California Trail. These were the routes of the covered wagons, the stagecoaches, and the mail-carrying pony express. In time these roads were improved by short cuts, bridging, and tunneling.

Eventually explorer trails became the routes for the transcontinental railroads. The Union Pacific and Central Pacific joined the East and the West along routes mapped by Jedediah Smith, Frémont, and the great surveys which followed the Civil War. The Santa Fe Railroad was built approximately along the line of the old Santa Fe Trail.

Attempts to unify the wagon roads of America into a national network made some headway in the nineteenth century. Slowly the federal government was able to get the separate states to join in linking together what was called the National Road, reaching as far west as the Mississippi.

It was not until the early decades of this century, the automobile age, that cross-country travel on paved roads became possible. In the summer of 1916 a dust-caked roadster came chugging into the city of San Francisco. The driver had just finished a week-long transcontinental trip. His route was the newly completed Lincoln Highway, the first modern road spanning the nation.

The Michigan Road, built to connect frontier settlements, became an important midwestern road network. Highway No. 1 runs from the Maine Coast, where Champlain once explored, through a sandy spit of land discovered by Juan Ponce de Leon —now called Cape Kennedy. The old Camino Real became the main coastal road of California.

Within the next few years this nation will complete its 42,500-

mile Interstate Highway System. Each year new maps add sections to this network of broad, high-speed, nonstop highways. Many follow ancient trails tramped out by buffalo on their way to salt licks or by explorers finding their way across the wilderness.

How many of the fast-moving travelers on Montana's interstate expressway will realize that they are following in the footsteps of Lewis and Clark? Will those journeying in Wisconsin on Interstate Highway 94 pause as they cross the portage path where Jolliet trudged on his way to the Mississippi? And will motorists racing through the Southwest know that Coronado once passed their way, searching anxiously for the golden cities of Cíbola?

Today's maps are guiding men across the skies—some piloting huge jet transports, others flying their small craft visually by means of aeronautical charts.

In an air-borne and space-borne age this land has taken on a more familiar appearance. Seen from greater and greater heights, the face of America comes into clearer view as the portrait which the mapmaker has been drawing these many years.

On a cold, clear morning, a helicopter hovers over the Sierra Nevada like a brown spider on an invisible web. Below are the rugged and wrinkled uplands Jedediah Smith once explored. The region is being mapped again—this time by the methods of modern cartography. The three men in the helicopter—the pilot, the engineer, and his assistant—are members of a large mapmaking team. Poised in mid-air, they are at one point of a triangular survey.

Below, far to the left, is the group of men who have been dropped off by the helicopter on a high ridge. There they have set up an observing station using light and radio beams. Below to the right, another party has come by truck through a heavily forested area to form the third point in the triangle. From the truck body

In the air-borne control survey system, angle and distance measurements are made from two ground stations to a helicopter hovering over the target point.

a high tower has been lifted above the treetops. On its platform the men have an unobstructed line of sight to the helicopter and to the party on the mountaintop. The survey observations are carried out by means of precision instruments that accurately measure angles and distances, both horizontal and vertical. The three groups carry on a lively conversation by radio.

When they all return to their base headquarters, another phase of the modern mapmaking operation will begin. Their data will be checked again and again. The necessary calculations will be carried out with the help of a computer.

Aerial photographs taken at high altitudes will be used to fill in the details of the map—relief features, vegetation, drainage, land forms, man-made objects. The photos are "read" by means of stereoscopic instruments which show up the terrain sharply. The pictures are studied in overlapping pairs, with the left eye looking at one picture and the right eye at another, the entire scene appearing in three dimensions.

The final map will resemble the photographs, but with differences. Unimportant and confusing details will be dropped out in the final drawing. However, the cartographer will add his observations of the exact location of the surveyed region in terms of the grid system of latitude and longitude. Altitudes above sea level will be recorded. The shape of the land will be traced by means of contour lines showing the levels of elevation.

In words, the finished map will show the names of mountains,

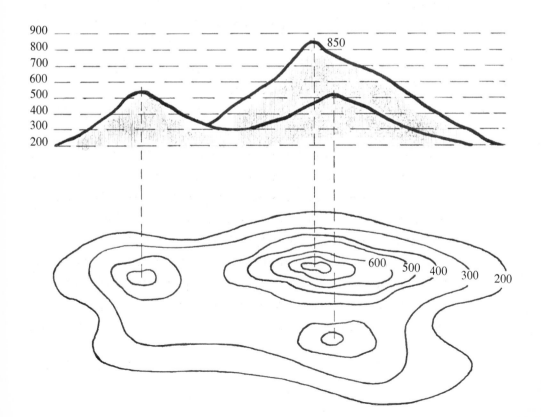

rivers, valleys, and other natural features, as well as such man-made structures as roads, bridges, dams, bench marks, fire towers, boundary lines. By means of symbols in color, the map will distinguish vegetation, buildings, mines, camp sites, towns.

The widely used aerial camera has made the cartographer's job speedier and more accurate. Photographs take the place of numerous handmade sketches and detailed notes. The scientific skills of photogrammetry are constantly being improved so that the camera records a rich pattern of information about the land. Photographs of the same area, taken periodically, clearly show the natural and man-made changes.

The 1960s put the cartographic camera in space satellites high over the earth. In the 1970s the "eye in the sky" program of mapping the earth from orbiting spacecraft is expected to open a new era of cartography.

The exciting possibilities are seen in some of the experimental photographs already taken. In the spring of 1968 the Apollo 6 satellite focused its camera on the Texas region where the cities of Dallas and Fort Worth are located. The result was a remarkable photomap in great detail. This map disclosed the amazing changes that had taken place during the sixteen years since the area had been previously mapped.

Gemini photomaps of the entire Southwest of the United States covered the "Coronado country," revealing topographic features that had never before been seen so clearly. These maps showed the usefulness of a permanent satellite which would uniformly repeat the mapping of areas after short intervals of time. By comparing such photomaps we could learn about the changes on the landscape and the rate at which they are occurring.

Experiments with infrared photography have proved that moisture in the soil can be detected. Regions with more moisture show up as "cooler" on such photographs. Maps put together from photographs of this kind can give us highly valuable information about use of land and dangers to crops, livestock, and wild life, and can suggest the necessary steps to be taken toward a better balance of water resources.

Satellites, cameras, computers, electronic sensors, high-speed color presses—these are all part of cartography today. And yet the "gadgets" have not replaced the human mapmaker. They have in fact made him more important, helping him to produce greater quantities and varieties of maps, a thousandfold more useful than before.

Aerial photographs still need to be checked by men in the field. The mapmaker must make on-the-spot observations so that he can understand, verify, and evaluate what the photographs show. Human skills are needed to devise the cartographic spacecraft and the instruments they carry. When computers are used in mapping, men must program them so that they will produce the kind of information that is desired.

The business of the mapmaker has always been to gather and sort facts and to produce a clear drawing of his information. Modern techniques and inventions are helping him in these ancient tasks. But the new technology has not solved all his problems. Mapping from aircraft and spacecraft is often hampered by clouds, fog, and mist. In a sense these have always been the mapmaker's problems. For centuries he has tried to see clearly through the shroud of ignorance and confusion. His has been a fight for the truth about the earth's surface. And like the men who went before him, today's mapmaker is still out on the frontier between the known and the unknown.

If we were to read between the lines and beyond the symbols of a map, to follow the windings of its rivers, the rise and fall of the terrain, the tracings of impassable wilderness, we would understand the problem of that map's makers. To look carefully at any map is to see the men behind it.

Today's map is the product of what men have achieved in centuries, often through high courage and great hardship. They have known thirst on the deserts and numbing cold on the lonely summits. Their tasks have sent them wading through dangerous swamps and plumbing the offshore depths of the sea. Many of them ended a wearying day clinging to some cliffside in the night, trying to sight on a star.

In such ways these men left their marks on our maps. They are honored in turn on today's maps by landmarks which bear their names. The Verrazano-Narrows Bridge is a spectacular span across the neck of New York Bay which the explorer Verrazano mapped. Lake Champlain recalls the mapmaker who discovered the valley northeast of the Adirondack Mountains and its large and placid lake. Towering over the Wind River in Wyoming is a peak called Fremont. The names of Lewis and Clark are repeated across the Western landscape on a mountain range, a national forest, and a narrow pass through the Great Divide.

These are some of the mapmakers who are now honored by enduring monuments. Along with many others, they mapped this country. Together they drew the likeness of our land.

Suggestions for
Further Reading

Suggestions for
Further Reading

Bakeless, Katherine and John. *They Saw America First*. Philadelphia: Lippincott, 1957.
The authors provide an easy-to-read review of exploration in America from Columbus to Lewis and Clark.

Bishop, Morris. *Champlain, The Life of Fortitude*. New York: Knopf, 1948.
Samuel Champlain's maps made America known to Europe, and this account tells of the man's courage in making a firsthand record of the wilderness of New France.

Bolton, Herbert Eugene. *Coronado, Knight of Pueblos and Plains*. New York: Whittlesey, 1949 (Division of McGraw-Hill).
Fifty years after Columbus, Coronado explored and mapped his way into what is now Kansas. This book is an account of his remarkable travels.

Brown, Lloyd. *Mapmaking, The Art That Became a Science*. Boston: Little, Brown & Co., 1960.
An outstanding cartographer explains the uses and the history of maps and the stories of their makers.

Darrah, William Culp. *Powell of the Colorado*. Princeton: Princeton University Press, 1951.
This author gives the full, stirring account of Powell's Colorado River trip, but also includes Powell's later work in developing a complete map of the United States.

Eifert, Virginia S. *Louis Jolliet*. New York: Dodd, Mead, 1961.
This is a well-written story of America's first native-born mapmaker and his historic Mississippi voyage.

Greenhood, David. *Mapping*. Chicago: University of Chicago Press, 1944.
Every aspect of mapmaking is explained in this lucid and well-illustrated book. It presents the art of cartography in simple language.

Morgan, Dale L. *Jedediah Smith*. Indianapolis: Bobbs-Merrill, 1953.
True-to-life adventure is the background of this narrative of a fur trapper's part in the opening and mapping of the West.

Powell, John Wesley. *The Exploration of the Colorado River*. Chicago: University of Chicago Press, 1957.
For those who would like to read Powell's own record of his stirring adventure, this is a very exciting diary.

Thomas, Dana Lee. *The Story of American Statehood*. New York: Funk & Wagnalls, 1961.
After the thirteen original colonies formed the United States, thirty-seven additional states joined the Union. This book tells of the men and events that shaped these states.

Tomkins, Calvin. *The Lewis and Clark Trail*. New York: Harper, 1965.
With good illustrations and maps, this concise little book takes us on the path that Lewis and Clark mapped to the Pacific.

Index

Index